THE PENINSULAR RANGES

A GEOLOGICAL GUIDE TO SAN DIEGO'S BACK COUNTRY

Michael J. Walawender
San Diego State University

KENDALL/HUNT PUBLISHING COMPANY
4050 Westmark Drive   Dubuque, Iowa 52002

Copyright © 2000 by Kendall/Hunt Publishing Company

ISBN 0-7872-6440-7

Printed in the United States of America
10   9   8   7   6   5   4   3   2   1

CONTENTS

LIST OF ILLUSTRATIONS

PREFACE

To the Reader

*M*any friends and acquaintances have told me that they took a geology class in college to fulfill their science requirement. They often express regret for not having taken the class more seriously and say that they would like to learn more about the rocks and landscapes that are now part of their lives here in the greater San Diego area. Others, in the course of their travels, have driven past rock outcrops with strange patterns of color and form or walked along high mountain ridges that look down on rolling landscapes and simply wondered how such things came to be. This book was written for all of you.

The geological history of greater San Diego is a story of gentle seas giving way to volcanic convulsions, of mountains rising and disappearing in the twinkling of a geologic eye, of churning magmas cooling fully six miles below an ancient surface to form gemstones with a hidden trail of gold dust. For geologists, the challenge is to look at scattered exposures of different rocks and, with some help from modern analytical tools, to reconstruct events that occurred hundreds of millions of years ago, and at depths and scales beyond human experience. The reader can drive through our magnificent back country to view the outcropping remnants of these events, and, along the way, travel through more than two hundred million years of geologic time.

I have also tried to make this book interesting to geologists who come to visit the greater San Diego area. San Diego's back country is a wonderful kaleidoscope of granitic rocks, awesome scenery, migmatites, gold mines, and pegmatites. To the south lies the charm and adventure of Baja. Although this book does not cover specific locales below the border, the rocks and geological relationships described ignore the political boundary and extend southward through the peninsula.

San Diego County covers 4,255 square miles. It ranks as one of the largest counties in the United States and is larger than the combined area of the two smallest states, Delaware and Rhode Island. An extensive road network, both paved and unpaved, allows access to all but the most remote terrains. This book is designed to take advantage of that access by providing maps and directions to key outcrops where the readers can observe first-hand the rocks and geological relationships. Nine field excursions and nearly thirty stops are included. Some stops include short walks along either the road, well established hiking trails, or, for the more adventuresome, through the infamous

California chaparral. Care was taken in selecting these sites so that each provides the optimal combination of safety, accessibility, and interest. At selected sites, you are taken through the logic used to make the critical interpretations about the history and origin of the rocks and features. In some ways, perhaps, these are detective stories with missing mountains, hidden treasures, and surprise endings. Enjoy.

 Acknowledgments

The author gratefully acknowledges aid and information from many sources. First, to my wife, Sonja, for her continuous encouragement and many careful readings of the manuscript. To Harl Hoppler for reading and editing the manuscript. To the countless alumni of the Department of Geological Sciences at San Diego State University, without whose theses and field projects, this book would not have been possible. To my colleagues, specifically, Gary Girty, Tom Rockwell, Gordon Gastil, Tom Demeree, and Pat Abbott, for sharing their knowledge and ideas. To Rene Wagemaker who drafted one of the more difficult figures. And finally, to the many fine people in the greater San Diego area that have, over the years, allowed me access to their land and mining claims.

AN OVERVIEW

 Science and the Study of the Earth

arl Sagan once observed that "Science . . . is a way of thinking." It is not a body of data or a set of theories and hypotheses. It is instead a philosophy of data acquisition and interpretation followed by more data acquisition, more interpretation, more data, and so on, that leads to the solution of scientific and societal problems. As these problems are defined, hypotheses are advanced and systematic data collecting programs are proposed. As data are collected and analyzed, ideas emerge and theories[1] are generated. Theories are then tested with new data sets and modified to fit the expanding data base. Ideas shown to be inconsistent with the enlarged data base are scrapped and new ones proposed. This is the critical aspect of scientific inquiry, i.e., the philosophy that it's acceptable, even admirable, to admit that the initial explanations were incorrect and to move on to new ideas and new horizons. Science learns from its mistakes and admits its own fallibility.

The advent of modern analytical tools has led to rapid changes in our understanding of how the physical world and the universe operate. One of the critical problems with regard to science and scientists is that the general public is deluged with stories in the popular press of great technology-driven scientific breakthroughs only to learn later that further studies have opened questions about the initial reports. Cold fusion and Martian microbes are part of that pattern. Skepticism of science and its proponents waxes and wanes accordingly but perhaps some measure of understanding should be applied to the process. Scientists are passionate about their work and may be driven into premature disclosure by fierce competition for research funds coupled with a bit of egotism. Scientists are, after all, still people. Good science, however, uses a system of checks and balances that involves peer reviews for papers and funding proposals. Although not a guarantee against inaccurate results or unnecessary research, such methodology tends to put academic brakes on poorly conceived studies before they find their way into the public realm.

---

[1]Theories and hypotheses are somewhat different. A hypothesis is an idea without supportive data, whereas a theory is a modified hypothesis that is supported by at least some hard data.

Geology as a scientific discipline is grounded in all of the fundamental sciences such as mathematics, physics, chemistry, and biology. However, it has its own paradigms, such as plate tectonics, that set it apart from the other "hard" sciences, and its own integrated concept of time, order, and scale that makes geologic thinking unique. Physicists and chemists understand the complex order of solid matter in relatively confined systems. Geologists must make the transition from order to disorder and back again during, for example, the creation of igneous and metamorphic rocks, and must conceptualize processes that run the gamut from atomic to global scales. Biologists delve into the nature of life and its origins. Geologists, specifically paleontologists, carry those concepts back nearly three billion years to study the evolution and diversity of ancient life forms. Geology is perhaps most closely aligned with astronomy. Astronomers must think in terms of vast distances and enormous spans of time that take us back to the dawn of creation. Geology requires the understanding of only a *few* billion years and distances limited to the confines of the planet. Like astronomers, geologists must work in four dimensions, the two quasi-horizontal dimensions of the Earth's surface, a third that projects us into the Earth's interior, and, finally, in time. Time, as hidden in the subtle arrangement of atoms and minerals in rocks, may extend back a few hundred million years or perhaps even billions of years. Each rock, like light from a distant star, contains the evidence of its history. Its birth, its transformations, its movements, and even its demise are there for us to decipher. A single bit of solid rock can record the uplift of mountain ranges, the motion of continents, the destruction of ecosystems, or the presence of materials so necessary to a technological society. It is this record in the Peninsular Ranges of San Diego County that is the focus of this book.

△ Southern California and the Pacific Southwest

The geologic features of Southern California are young relative to the rest of the North American continent. Californians live and work on a portion of the Earth's crust whose history goes back only a few hundred million years. This stands in stark contrast to the ancient rocks and prolonged history found a short distance to the east in Arizona where exposed igneous rocks are as old as 1.8 billion years (1800 million years). The rocks found in Southern California and Baja California have not been subjected to the multitude of geological events such as mountain building, volcanism, and erosion that affected the former western edge of the North American continent over the previous 1500 million years. Instead, they provide a marvelous record of how new material, from the massive sedimentary rocks of the Great Valley to the towering granite monoliths of the Sierra Nevada, was created and added to the North American continent. This record has been distorted somewhat by geologically recent movements along the San Andreas and related faults, but can be reconstructed to reveal a broad continuity of processes that began in Jurassic time (Figure 1.1) and continued unabated for more than 100 million years.

Figure 1.2 shows a generalized physiographic map of Southern California. Each physiographic province contains rocks that are associated in space and time and that lie within a recognizable physical setting. San Diego County includes two of these provinces, the Peninsular Ranges and the Continental

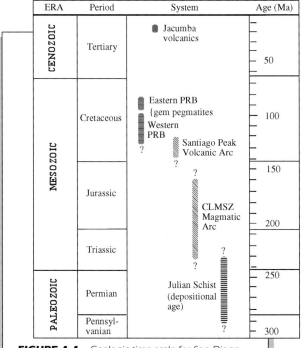

ERA	Period	System	Age (Ma)
CENOZOIC	Tertiary	● Jacumba volcanics	50
MESOZOIC	Cretaceous	Eastern PRB {gem pegmatites Western PRB — Santiago Peak Volcanic Arc	100
MESOZOIC	Jurassic	? ? — CLMSZ Magmatic Arc	150 — 200
MESOZOIC	Triassic	? ?	
PALEOZOIC	Permian	Julian Schist (depositional age)	250
PALEOZOIC	Pennsylvanian	?	300

FIGURE 1.1 *Geologic time scale for San Diego County. Bars represent the time span for the geologic events discussed in this text.*

Borderland, and borders a third, the Salton Trough. Most of the Continental Borderland lies offshore where it contains numerous northwest-oriented faults that, along with the San Andreas to the east, continue to splinter much of Southern California and drive it northwestward into the Pacific Ocean. The rocks in this mostly submerged geomorphic province are roughly equivalent to those found in the Franciscan complex and the Great Valley Sequence. They, along with their age and spatial relationships with the Sierra Nevada, provide a geologic picture that can be extended southward to include Southern and Baja California.

In the context of plate tectonics, the Franciscan-Great Valley-Sierra Nevada triad represent the products created at the collision of two plates, one containing oceanic materials and the other a large continental mass (Figure 1.3). Rocks of the Franciscan terrane are described by geologists as a **melange**, a complex, physical mixture of sedimentary rocks that formed within the deep ocean basin and the quartz- and feldspar-rich sands and muds derived by erosion of a nearby continent. This mixing oc-

FIGURE 1.2 *Generalized physiographic provinces of Southern and Baja California. Salinia and Franciscan terranes are subdivisions of the Coast Ranges physiographic province and are separated here for clarity. The Sierra Nevada, Salinia, Transverse Ranges, and Peninsular Ranges contain the deeply eroded remnants of a single continuous volcano-plutonic arc that formed during the Mesozoic, whereas the Great Valley, Franciscan terrane, and Continental Borderland hold, in part, the sedimentary record of the uplift and erosion of that arc.*

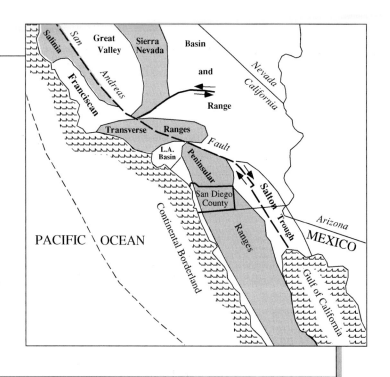

curs within a longitudinal trench that marks the zone of subduction in which the plate carrying oceanic materials is forced beneath the plate supporting the continent (Figure 1.3). Sands and muds eroded from continental source areas north of the cross section were moved slowly southward by ocean currents and plate motions into the trench. Chemically precipitated sediments such as chert and limestone that were deposited within the deep ocean basin west of the trench were carried by plate motions to the trench where they were scraped off and mixed with the continental debris coming in from the north.

With time, the underthrusting motion of the oceanic plate forced the leading (western) edge of the continental plate to rise slightly, creating a forearc basin between the continent and the trench. Weathered debris from the continent began to spill westward into the basin. As plate motions drove the subducted slab deeper and deeper into the hot interior, the oceanic crust reached its melting temperature and began to leak molten rock upwards to form a volcanic arc inland from the trench and forearc basin (Figure 1.3). As the magmatic arc rose to great heights above the surrounding terrane, the forces of erosion took hold and delivered weathered volcanic materials into the forearc basin to form the Great Valley Sequence. These complex processes will be treated further in subsequent chapters but the net result was the creation of a lateral succession of geologic environments ranging from the trench-related melange to the relatively quiet sedimentation within the forearc basin and finally to the emerging volcanic arc whose plutonic roots now form the Sierra Nevada.

This tectonic and physiographic system extended unbroken from Northern California southward into what is now Baja California until about 20 million years ago when lateral movement along the San Andreas Fault system began to slice the components into smaller pieces and rotate them into their present configuration (Figure 1.2). The Salinian Complex, the Transverse Ranges, the Peninsular Ranges and the Sierra Nevada all, at one time, formed a more or

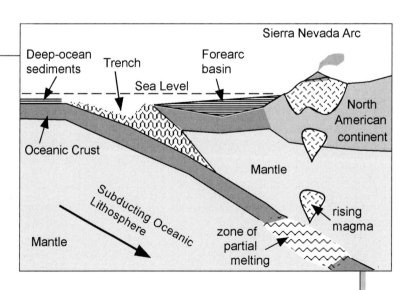

FIGURE 1.3 Plate configuration during the Cretaceous. Oceanic lithosphere, consisting of deep ocean sediments deposited on top of basaltic oceanic crust and its underlying mantle peridotite (see Chapter 3), is forced beneath the western (leading) edge of the North American continent. The Franciscan melange, a mixture of those deep ocean sediments and loose material derived from the continent, was being deposited in the offshore trench that marks the surface boundary between the two colliding plates. The massive Great Valley sediments were accumulating in the forearc basin just west of the Sierran arc.

less continuous volcanic-plutonic magmatic arc westward from which lay the forearc Great Valley and trench-related Franciscan sediment traps. For us in San Diego County, these two sedimentary basins are mostly hidden beneath the ocean. Bits and pieces are exposed in offshore islands such as Santa Catalina and San Clemente. Some of the Great Valley sedimentary rocks are exposed in coastal La Jolla and Point Loma.

The events discussed in this book will range back to a time before the onset of subduction and carry forward to cover the evolution of the Peninsular Ranges and the younger volcanic events that pierced the granitic rocks in central San Diego County, a span of more than 200 million years. Many concepts will be introduced but none more important than the measurement of geologic time. Several techniques for measuring the relative and absolute ages of rock units will be discussed. **Relative time**, the ordering of geologic events without regard to numerical age, is determined from the geometric patterns between adjacent rock units in the field. **Absolute age** refers to the numerical age of the rock and is the time in the past when the rock was formed. Geologists use "**Ma**" or "**Ba**" as abbreviations for millions or billions of years before present, respectively. Highly trained technicians using sophisticated, expensive, and complex analytical tools such as mass spectrometers carefully measure a variety of radiogenic elements in the individual minerals of a rock, and from this calculate absolute ages. We begin by discussing the most commonly used technique for determining the absolute age of igneous rocks

△ Radiometric Age Measurements

Although the oldest rock yet dated from the Earth's surface is only 4.0 billion years old, the Earth itself is nearly 4.6 billion years old. That latter number was not easily obtained and is the product of separate lines of evidence that include radiometric age measurements on lunar and meteorite samples. It reflects the enormity of time that has passed since the Earth accreted from a spinning mass of galactic dust into a coherent body with a more or less solid outer shell.

Early attempts at estimating the age of the Earth were usually based on tenuous scientific grounds. Tracing the genealogical history of the biblical patriarchs as depicted in the Old Testament, James Ussher, a 17th century cleric and scholar, determined that the Earth was created in the year 4004 B.C. Could towering mountain ranges or vast ocean basins have been created in either a biblical instant or a few thousand years? Most scientists and even amateur naturalists considered 6000 years to be too scant a time period to account for the variety of features found on the Earth's surface and sought techniques by which the Earth's age could be established. Charles Darwin had suggested that at least 200 million years was needed to account for the complexity and diversity of the life forms he had observed. The rate at which sediments accumulate in ocean basins was compared to the total thickness of known sedimentary rock sections and yielded ages between 3 million and 1.5 billion years. The rate of salt accumulation in the oceans was back-calculated to the point where the oceans were composed entirely of fresh water and yielded an age of about 90 million years.

The British physicist, Lord Kelvin, used the cooling rate of the Earth to calculate that it could not be older than 100 million years. Thus, by the early part of the twentieth century, most scientists accepted the age of the Earth to be on the order of some tens of million of years, a number greatly different from both the calculated age of Bishop Ussher and the presently accepted value of nearly 4600 million years. In 1902, Ernest Rutherford, a British physicist, published his *Theory of Atomic Transmutation* in which he argued that radioactive elements spontaneously emit charged particles from their nuclei. This transforms the original (parent) atoms into new (daughter) elements. By 1910, geoscientists at Yale University had applied this principle of radioactive decay and determined rock ages up to 1600 million years from three different continents. These large numbers were not readily accepted by the scientific community and it would not be until the 1950s that radiometric dating techniques would win world-wide acceptance. The idea of radioactive decay, in its simplest form, argues that the isotopes of certain elements such as uranium (U) spontaneously emit charged particles at a constant rate and, in doing so, change to a different element such as lead (Pb) with lower mass. This process is random at the level of the individual atom so that in a given population of the same radioactive elements, the decay rate is measured in terms of the amount of time necessary for one-half of the existing parent atoms to decay to the daughter element. Thus, if the measured half-life for a given radioactive element is 10 million years and if we started with N atoms of the parent element, N/2 atoms will remain after 10 million years. After 20 million years or two half-lives, one-fourth of those atoms ($1/2 \times N/2$) will remain, and so on. So, the time at which the radiometric clock was turned on (the age of crystallization of an igneous rock) can be determined by measuring the abundance ratio of the daughter element to its parent and applying that to the known rate of decay (Figure 1.4).

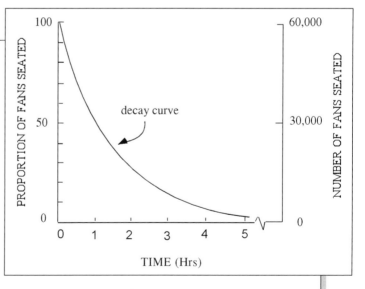

FIGURE 1.4   Imagine that we fill a Padre game at San Diego's Qualcomm Stadium with 60,000 fans (N_0) and that one-half of the seats are computerized to fold up at random during each one-hour period (one half-life) and force the occupants to stand. No one is allowed to leave. The game goes into extra innings but at 5:15 P.M., Tony Gwynn doubles in the winning run. At that exact moment, the computer tells us that there are still 3,750 fans seated (N_t). At what time did the game start? In this example, the start of the game represents the time in the past when the "rock" formed, the 3,750 fans still seated at the end represent the remaining original "atoms," and the 56,250 standing fans represent the newly created "daughter elements." As shown in the diagram, after one hour (one half-life), 30,000 fans are still seated; after two hours 15,000, after three hours 7,500 and, after four hours, 3,750. It took exactly four half-lives to reach this point (5:15 P.M.) so the game was exactly four hours old and started at 1:15.

In order to apply U-Pb geochronology to the measurement of the age of an igneous rock, a specific mineral must be extracted from the rock which not only contains the parent (uranium) and daughter (lead) elements but from which these elements cannot have escaped over the course of geologic time. The measured age gives the time in the geologic past when the mineral (and thus the rock) crystallized. Zircon, a zirconium silicate that contains small amounts of uranium, is found in most igneous rocks but in very minute quantities. Typically, 100–200 pounds of fresh rock are collected, crushed, and passed through a fine-mesh filter. The homogeneous powder is washed carefully on a table-like device similar to the sluice box used in placer gold mining. The lighter, abundant minerals such as quartz and feldspar are washed away. Zircon and other heavy minerals from the rock are recovered and then separated by floating and sinking them in selected high-density liquids. The zircon-rich residue is examined microscopically and individual zircon crystals are hand picked for dissolution in acid. The acid solution goes into a mass spectrometer which measures the abundance of all the isotopes of uranium and lead. Knowing the rate of decay of a given isotope of uranium to its specific isotope of lead, it is, in theory, a simple calculation to determine the age in the past when the zircon (and the rock) solidified. Analytical errors are typically on the order of one percent so that the reported ages of 100 Ma will carry an uncertainty of about 1 million years.

Zircon U-Pb geochronology has been used almost exclusively to determine the age of emplacement and crystallization of the igneous rocks within San Diego County. These ages vary from over 200 million years ago (Ma) to less than 90 Ma and document a complex pattern of intrusion, deformation, and uplift that produced the topography so familiar to us today.

The next few chapters will describe the sequence of events that created the oldest rock units in our area. These events range from quiet sedimentation in shallow seas adjacent to the North American continent to collisions between lithospheric plates that produced the three separate volcanic arcs and their plutonic counterparts listed in Figure 1.1. Uplift and erosion have reduced the earliest of these rock masses to mere shadows of their former selves and have stripped away the volcanic cover of the youngest to reveal its plutonic roots. These younger rocks are the most abundant and form the Cretaceous **Peninsular Ranges Batholith (PRB)**, a sequence of igneous rocks that extends from Riverside County southward through the State of Baja California. It has dominated the topography throughout the county for tens of millions of years and is responsible for many of its familiar landmarks. The early geologic history of what is now San Diego County, however, begins sometime prior to early Jurassic.

CHAPTER 2

THE OLDEST ROCKS

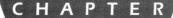 ## The Triassic (?) Julian Schist

Geology

*T*he Julian Schist is named for a series of metamorphosed sedimentary rocks first studied in the Julian area by F. S. Hudson in 1922. The name has since been applied to essentially all of the ancient metasedimentary rocks found within the Peninsular Ranges Batholith. Outcrops of these rocks occur throughout San Diego, Imperial, and Riverside Counties and extend southward through the State of Baja California. Their distribution (Figure 2.1) is somewhat spotty because they represent the remnants of an extensive package of ancient sedimentary rocks that have been disrupted and metamorphosed during the emplacement of younger plutonic rocks. Approximately ten kilometers of erosion since Mesozoic time has further reduced their presence to isolated **roof pendants**, packages of metasedimentary rocks entirely enclosed by younger plutonic rocks. The story of these oldest known rocks within the PRB has been pieced together by studying the scattered exposures one by one.

Schist is a term for a metamorphic rock in which there are abundant platy minerals such as biotite and muscovite. These minerals are aligned in a common direction giving the rock a layered or leafy appearance similar to the pages of a book. Within the Julian Schist, this term is somewhat of a misnomer since few of the exposed rocks are indeed schistose. In fact, these rocks exhibit a lateral variation in composition as one travels from west to east. In the western and central portions of the county, the Julian Schist is mainly a biotite-bearing quartzite (see Appendix B) with lesser amounts of biotite schist and quartz-pebble metaconglomerate. As one proceeds eastward, more and more schistose rocks, some muscovite-bearing, appear and form a thick sequence with alternating layers of schist and quartzite. Still further to the east near the San Diego-Imperial County line, the character of these metasedimentary rocks changes abruptly. The amount of schistose rocks has decreased, the quartzites contain little or no biotite or muscovite, and marbles now form a significant portion of this older rock assemblage.

This eastward change in rock type is accompanied by an increase in the degree of metamorphism. During metamorphism, the original materials in the parent sedimentary rock react with one another to create a new set of miner-

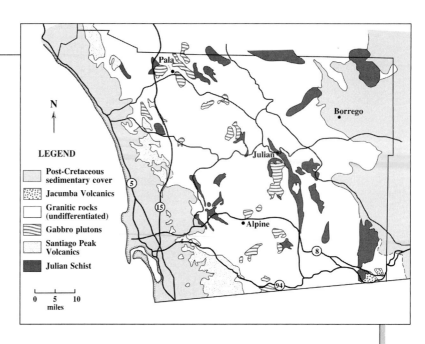

FIGURE 2.1 *Generalized geologic map of San Diego County with the major rock subdivisions listed from youngest to oldest in the legend. The light shaded regions represent the younger (post-Cretaceous) sedimentary cover over the Peninsular Ranges Batholith (PRB). To the west, this cover lies within the continental borderland and, to the east, along the eroded margin of the Salton Trough. The unpatterned zone (granitic rocks) and striped patterns (gabbroic rocks) within it constitute the PRB. The Santiago Peak Volcanics (stippled pattern) and the Julian Schist (dark shaded areas) represent the remnants of the volcanic and sedimentary cover, respectively, into which the plutonic rocks of the batholith were intruded.*

als (see Stop 2-1) that are characteristic of the pressures and temperatures at which metamorphism occurred. This process takes place at temperatures and pressures below the melting point of the rocks and *occurs entirely in the solid state*, i.e., the rocks do not melt. In the west, metamorphic minerals such as andalusite and cordierite suggest that metamorphic temperatures and pressures were near 600°C and 2.5 kb, respectively. In the east, minerals such as silli-manite, diopside, and wollastonite indicate about 640°C and 3.5 kb. The pressure on a metamorphic rock during metamorphism is due to the depth of burial, i.e., the weight of the overlying column of rock. Using the general conversion factor for crustal rocks of 1 kb = 3.5 km (2.1 miles), pressures of 2.5 and 3.5 kb indicate a burial depth of about 9 km (5.5 miles) in the west and 12 km (7.3 miles) in the east. *Since these rock are now exposed at the surface, that thickness of rock must have been stripped off the top of the PRB since the time of metamorphism.* It also indicates that the rocks in the east were buried more deeply than those in the west, a point that we will return to in a later section.

Finally, scattered throughout all portions of the Julian Schist are isolated exposures of dark, fine-grained amphibolite, a metamorphic rock composed of hornblende and plagioclase. Recent studies have shown that these rocks are the metamorphosed equivalent of basalt and that they represent lava flows, dikes, or sills within the Julian Schist. Basaltic lavas form in only three general tectonic settings:

1. mid-ocean spreading centers
2. intra-plate or within-plate hot spots such as Hawaii or Yellowstone National Park
3. volcanic island arcs in the early stages of their development.

The chemistry of these metamorphosed basalts is distinctive and can be correlated to basaltic rocks erupted at mid-ocean spreading centers. Thus, it seems that the depositional basin for the sediments that eventually became the Julian Schist must, at some point, have endured a spreading or rifting event that created the intimate spatial association of sedimentary and basaltic igneous rocks. These basaltic rocks, now amphibolites, are the oldest known igneous rocks in our area and represent a tectonic setting quite different from and much in advance of that which produced the Peninsular Ranges batholith.

Depositional Environments

*D*espite the metamorphic overprint on these ancient sedimentary rocks, geologists can still determine the nature of the original sedimentary rocks (**protolith**) and make inferences about their paleoenvironments. The biotite-bearing quartzites and minor schistose rocks in the west represent quartz-rich sands and clay-rich muds, and suggest deposition of sediments into a shallow sea adjacent to a stable continental margin, a setting not unlike that of the modern-day east coast of the United States. The rocks still have recognizable shallow-water sedimentary structures such as cross beds and ripple marks. Microscopic examination reveals that the original quartz grains were well rounded and nearly spherical, a shape that can only be created through rigorous transportation of the grains in river channels, beach environments, or both. The interbedded character of the quartzites (sands) and schists (muds) suggests that this environment was subjected to periodic rise and fall of sea level so that sands would migrate seaward to overlie shales during periods of **regression** (falling sea level) with the reverse true during periods of **transgression** (rising sea level). To the east, the marble-quartzite assemblages also suggest deposition in a shallow, continent-fringing sea but with a distinct change in depositional style. The pure quartzites argue for an environment in which the sands were washed clean of most of their mud fraction, something equivalent to active beaches, whereas the marbles represent mostly silt-free limestones that form in warm shallow seas.

Age

*D*espite the striking similarity in depositional environments, one problem remains: the age of the Julian Schist. In an undisturbed sequence of layered sedimentary rocks such as those exposed in the beach cliffs throughout the San Diego area, the rocks get younger as one goes upward from the base of the exposure. Thus, a sandstone bed that overlies a mudstone layer is younger because it was deposited on top of the existing mudstone, and so on. In tilted sedimentary rocks, however, the "up" direction is difficult to recognize so that relative ages cannot always be ascertained. The exposures of Julian Schist everywhere show evidence of at least two periods of deformation with the rocks now bent into a series of tight folds with steeply inclined axes. In most exposures, the beds are steeply dipping to vertical so that the "up" direction is difficult to determine. For outcrops separated by even modest distances, the relative ages cannot be ascertained. Thus, it is not known whether the quartzite-marble sequences in the eastern zone are older or younger than the quartzite-schist sequences to the west and they have all been lumped into one prebatholithic age category.

Only one fossil has been reported within the Julian Schist and it has somehow been lost to science. The fossil was described by F. S. Hudson in 1922 as a cast of an ammonite shell with a "probable" Triassic age. Despite the uncertainty associated with that determination, a Triassic (248 to 206 Ma) depositional age for the Julian Schist is consistent with its observed relationships to the igneous rocks in the county. The oldest plutonic rocks are dated at 234 Ma. They contain fist-sized and larger fragments of the Julian Schist that had fallen into the invading magma and must therefore be older than (were deposited before) the enclosing igneous rock (Stop 1-2, Figure 2.5).

Another method used to estimate the age of the Julian Schist is to compare the rock types and their inferred depositional environments to packages of sedimentary rocks exposed elsewhere on the North American continent. The rocks that make up the western Julian Schist are similar to the Barranca Formation, a group of Triassic sedimentary rocks found in northern mainland Mexico. If the Julian Schist does in fact represent a northern and western extension of that sedimentary system, then the reported fossil age appears to be reasonable. The quartzite-marble assemblages, however, are similar to rocks of Early to Medial Paleozoic age (see inside front cover) found throughout the Southwestern U.S., and Ordovician fossils have been extracted from marbles in the Coyote Mountains east of Ocotillo, California. Although the rocks in the Coyote Mountains are separated from the bulk of the quartzite-marble sequence by the left-lateral Elsinore Fault and may have been tectonically transported into the area by movement along the fault, similarities between the two quartzite-marble sequences are striking. If this correlation is correct and the quartzite-marble sequence is Paleozoic in age, then the two rock sequences that make up the Julian Schist are distinct sedimentary systems, separated in depositional age by hundreds of millions of years, that have been juxtaposed by the tectonic events that shaped San Diego County. However, one critical and unifying factor remains: both rock sequences are made up of sediments derived from continental sources and deposited into shallow seas.

Summary

*D*espite the uncertainty in the age of the Julian Schist, its depositional setting is well understood. If we were to have an overview of this area in mid-Paleozoic to Triassic time, it would contrast greatly with what we recognize today. The landmass of the North American continent would be well to the east with Nevada, Washington, and Oregon covered by shallow seas. The ancient rocks found in what is now Arizona and Sonora, Mexico, sat as a modest highland shedding sediments westward. These sediments accumulated as layers of sand and mud and are the precursors to the western quartzite-schist sequence of the Julian Schist. They would continue to accumulate as sea level slowly rose and fell over millions of years. Snapshots in time would show rivers emptying their sediment loads into the shallow seas creating offshore sand bars, beach environments, small lagoons, but not great tidal flats or the warm seas that permit the accumulations of limestones or other rocks chemically or biologically precipitated from the ocean. As sea level rose in response to distant changes in plate motions, the muds and silts deposited far offshore would slowly migrate landward over the older near-shore environments and then back again to create the alternating sand-mud layers. Still later, behind us, to the west, we might see smoke from distant volcanoes of an emerging volcanic arc created as two plates collide even further to the west. Eventually, the calmness of the sedimentary basin would be shattered by rifting associated with the distant volcanic arc. This back-arc spreading center would erupt submarine basaltic lavas, now amphibolites, into and perhaps on top of the sand-mud layers. The San Diego topography so familiar to us today was still a long way into the future.

△ The Late Triassic-Jurassic Cuyamaca-Laguna Mountain Magmatic Arc

Geology

A dramatic change must have occurred over the great span of time between the deposition of sediments that were to become the Julian Schist and the next event observed in the geologic record of San Diego County. Although there is little known about this event, it must have involved plate collisions, subduction and the generation of an ancient volcanic arc whose presence is observed only in a relatively narrow belt of strongly deformed igneous rocks in the Laguna Mountains (Figure 2.2). This belt of deformed rocks, the Cuyamaca-Laguna Mountain fault zone (CLM), consists of relatively light-colored, gneissic igneous rocks that were intruded into the sedimentary precursors of the Julian Schist which then lay at the western edge of the North American continent. These gneissic rocks are typically biotite-bearing granodiorite and tonalite (see Appendix A), and are similar to some of the rocks found within the younger PRB. Geologists have long suspected that these gneissic rocks were older than the Cretaceous batholith that encloses them but proof of that relative age was lacking. Recently, U-Pb zircon ages from these deformed granitic rocks show them to be as old as 234 Ma and as young as 160 Ma. These ages are much older than any previously determined within the PRB (Figure 1.1). In addition, the rocks also contain a small population of zircons with an average age near 1400 Ma. Such a mixture of ages could only come about if the rising magmas intruded through the leading edge of the North American continent and incorporated fragments of those ancient rocks.

The Late Triassic to Jurassic magmatic arc rocks may have been much more extensive in this area than currently recognized and may have formed the crustal foundation on which successively younger magmatic arcs could develop. Similar granitic rocks of Jurassic age are found in a broad, discontinuous belt along the western side of the Sierra Nevada, but this belt thins considerably in San Diego County before turning southeastward towards what is now mainland Mexico. The rocks bear testimony to the existence of a larger Jurassic volcanic arc whose surficial manifestations have long since been stripped away, leaving behind only scattered outcrops of the arc's plutonic roots that were metamorphosed during the younger tectonic events which created the Sierra Nevada and Peninsular Ranges batholiths.

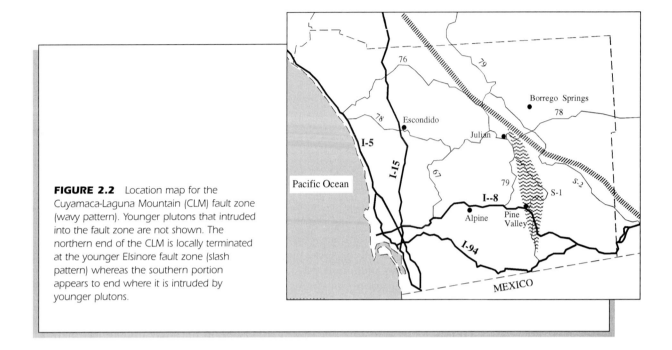

FIGURE 2.2 Location map for the Cuyamaca-Laguna Mountain (CLM) fault zone (wavy pattern). Younger plutons that intruded into the fault zone are not shown. The northern end of the CLM is locally terminated at the younger Elsinore fault zone (slash pattern) whereas the southern portion appears to end where it is intruded by younger plutons.

Even the ocean basins have not yielded a clear sedimentary record of the erosion and sedimentation that must have accompanied the growth of the CLM volcanic arc. Latest Jurassic sedimentary rocks are exposed in several coastal canyons in San Diego County. These rocks, however, were derived from a volcanic source terrane that was immediately adjacent to the sedimentary basin, not thirty or forty kilometers inland where the CLM is currently exposed. They do not contain any record of the plutonic rocks now exposed in the CLM leaving us to wonder and speculate. If we could go back in time, we might see a very different picture from today. The volcanic portion of this magmatic arc was built on or close to continental crust that was already above sea level, a setting that might compare to the modern Andes Mountains. The Andes, however, have a record of magmatism that goes back continuously for at least 150 million years. Our limited data on this volcanic arc suggest that it may have an age span of half that figure. Even so, it must have been an impressive mountain range that stood high above a coastal zone that collected its weathered residue. The sediments shed from this ancient mountain range have yet to be found. Are they buried somewhere beneath younger sedimentary rocks? Could our portion of the CLM belt have been rafted into its present position along now dead faults so that the sedimentary record of its volcanic system lies unrecognized somewhere else? Do more plutonic rocks of this age and composition exist further east in Imperial County? The search for these ancient rocks and their sedimentary residue continues but until more are found and studied, their role in the geologic evolution of the North American continent and San Diego County will remain a mystery.

Field Excursion 1

STOP 1-1

The Oldest Rocks

JULIAN SCHIST ALONG SUNRISE HIGHWAY. This field excursion can begin at either end of Sunrise Highway (S-1) that cuts across the backbone of the Laguna Mountains. The southern end is reached from I-8 about 35 miles east of San Diego and about 1.5 miles east of the exit for Pine Valley. To reach the northern terminus and its intersection with State Route 79, exit I-8 about 28 miles from San Diego at the Descanso off-ramp. Follow State Route 79 north about 17 miles through Cuyamaca Rancho State Park and past Lake Cuyamaca to its intersection with S-1.

Green mile markers are noted for this part of the trip log so that the stops can be accessed in either direction.

From its northern end, Sunrise Highway (S-1) has a series of road cuts in the Julian Schist starting about 2.6 miles southeast of its intersection with State Route 79 (Figure 2.3). About one-quarter mile further is a small road cut with limited parking on the east side of the road. The outcrops on the east side of the road are mainly amphibolite, a dark, dense hornblende-plagioclase metamorphic rock that represents one of the ancient basalt injections discussed in the text. Recent road work has obscured these outcrops so that the amphibolite is only exposed at the north end of the outcrop along the west side of the road. The remainder of the roadcuts consist of fine-grained mica schist and quartzite typical of this portion the Julian Schist. Ahead one-quarter mile is the parking area used for installing or removing tire chains when required in winter. The outcrops here are fairly weathered but consist of recognizable micaceous schist and minor quartzite.

Continue ahead (south) another one-half mile to high roadcuts on either side of the road. These, near mile marker 33.5, expose some of the best examples of the Julian Schist. Park on the left (east) side of the road. The outcrops consist mainly of very thinly bedded muscovite and biotite schists. The light-colored tabular body in the middle of the outcrop that cuts across the foliation is a narrow, fine-grained, garnet-bearing granite dike that is probably related to the pegmatite-aplite dikes east of this location (see Chapter 5).

An important aspect in understanding metamorphic rocks is the recognition of their **protoliths**, i.e., the original rocks that were metamorphosed to form the mica schists, quartzites, and so on. This can be done, in part, by determining the minerals that form the metamorphic rock, crudely estimating the bulk chemistry from the mineral compositions, and then finding a reasonable chemical match among other rock types. The micaceous schists, for example, contain abundant quartz, muscovite, and minor biotite. Sharp-eyed and experienced metamorphic petrologists might also notice faint, scattered clusters of tiny, white, lustrous fibers in the outcrop that are probably the mineral sillimanite. Muscovite contains abundant potassium, aluminum, and

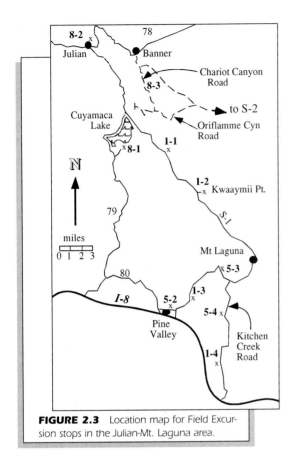

FIGURE 2.3 Location map for Field Excursion stops in the Julian-Mt. Laguna area.

silica. Sillimanite has only aluminum and silica whereas biotite is composed of potassium, aluminum, silica, and some iron. Since there is little element migration during metamorphism, the original rock (protolith) must also have been very rich in those chemical elements. For that crude chemical composition, the most probable protolith was a clay-rich mudstone, a type of sedimentary rock commonly formed offshore in moderately deep waters along the landward edge of an ocean basin. Quartzites consist mainly of quartz with small amounts of feldspar and likely represent coarser grained nearshore sediments washed clean of any mud by the motion of waves and tides. Thus, the metamorphic rocks exposed at this location represent sediments that accumulated at the shoreward fringe of an ocean basin. This sedimentary package was buried deep enough beneath the ocean basin to eventually lithify and undergo metamorphism, only to be uplifted through mountain-building processes and exposed again nearly 5,000 feet above sea level.

As you walk along the outcrop, notice that the rocks are steeply tilted. Since sedimentary rocks are deposited in more or less horizontal layers, these rocks have been tilted to their present position during an episode of mountain building (tectonism). As you trace your finger perpendicular to the layering, you cannot tell which direction represents the original "up" side of the layer but imagine that it is to your left. As you continue to trace your finger across the outcrop from the schist to a thin, blocky quartzite layer and then back again into the schist, you have gone up in time from a muddy sea floor into a beach or sand bar and then back into the deeper water where mud settles to the bottom. Why did these ancient environments change? Were the beach sands (quartzite) deposited during a brief lowering of sea level so that rivers flowing into the ocean basin could drop their sand-sized sediment further out to sea and atop the muds? Does the quartzite represent a 100- or 500-year storm discharge that pushed its sand load much further offshore? Either solution is compatible with surficial processes that are observed today.

CLM GRANITIC ROCKS ON KWAAYMII POINT ROAD. Kwaaymii Point, a jumping-off point for local hang-glider enthusiasts, is about six miles north of the town of Mt. Laguna and just north of mile marker 30 on Sunrise Highway (S-1). Turn southeast and drive about one-half mile to the end of the road (Figure 2.3). The wide trail that extends southward from the parking and view area is what remains of the old portion of Sunrise Highway that was repeatedly threatened by rock falls from the adjacent cliffs. The rocks exposed along the old roadbed are Jurassic granitic rocks caught up in the movement in the CLM fault zone. The predominant minerals are cream-colored feldspars, quartz, and dark biotite that have been strung out into a steeply dipping north-south foliation that crudely parallels the cliff face. This foliation plane is a zone of weakness in the rock along which the cliff face would fail causing the repeated small rock avalanches that made this stretch of road a headache to the county highway department. In the late 1970s, the road was shifted to its present course but one rockfall remains across the road about 1/4 mile south of the parking area (Figure 2.4).

 he view from the parking area is one of the most spectacular in the county. The green area below is Agua Caliente Springs and, just to its left (north), County Highway S-2 winds up Campbell Grade and into Box Canyon. The large prominent mountain beyond the townsite is Whale Peak, a large mass of 95 Ma granitic rock that is part of the eastern zone of the Peninsular Ranges Batholith (see Chapter 5). The active Elsinore fault zone runs along its southwest-facing slopes as it heads northwest towards Banner Grade and Volcan Mountain near Julian.

A large roadcut along Sunrise Highway (S-1) just a few yards north of its intersection with Kwaaymii Point Road exposes the same granitic rock described above. This outcrop contains large inclusions of micaceous schist and amphibolite (Figure 2.5) that resemble the Julian Schist (see Field Excursion Stop 1-1). These fragments are enclosed in a granite that may be as old as 234 Ma and indicate that the Julian Schist (see previous section) is at least that old.

STOP 1-3

CUYAMACA-LAGUNA MOUNTAIN (CLM) FAULT ZONE. If you started from the north end of Sunrise Highway, continue south another 15 miles (Figure 2.3). Otherwise, take the Sunrise Highway exit north from I-8 and drive 1.5 miles to a wide, flat pullout on the northwest side of the road (north of mile marker 15). Park here, cross the road, and walk north along Sunrise Highway about 100 yards to a large roadcut.

FIGURE 2.4 *Rockfall along abandoned stretch of Sunrise Highway (S-1) near Kwaaymii Point. The foliation (mineral alignment) in the rock is parallel to the rock surface and contributed to many small rockfalls along this section of the road.*

FIGURE 2.5 Fragment of dark amphibolite (below lens cap) and lighter colored Julian Schist (top center) within gneissic granodiorite near the intersection of Kwaaymii Point Road and Sunrise Highway (S-1). Two thin, light-colored dikes cut through the mineral alignment.

 Please exercise caution here as the shoulder is narrow and the motorists distracted.

The outcrop exposes a gneiss with a very pronounced foliation (mineral alignment) defined by elongate aggregates of biotite and feldspar similar to that seen at Stop 1-2. The original rock was a granodiorite (see Appendix A) but it has been sheared into a strongly foliated rock by movement in the fault zone. The foliation plane defined by the mineral alignment is lined up towards the northeast at about N53E. In order for the original igneous rock to have been recrystallized in such a plastic fashion, i.e., without breaking apart brittlely, the fault movement had to occur when the rocks were very hot and plastic but below their melting temperature.

This rock has a zircon U-Pb age of about 160 Ma which measures the time at which the pluton crystallized. It also contains an older population of zircons that have an average radiometric age close to 1400 Ma and indicates that the original magma must have risen through much older rocks that formed the western edge of the North American continent. Movement on the fault zone must be younger than either determination and can further be restricted to between 118 and 94 Ma. Across the road to the northwest lies the Pine Valley pluton (Stop 5-2). A portion of that body extends into the fault zone and has been deformed into a gneissic rock similar to that seen at this outcrop. Since that pluton has a reported zircon U-Pb age of 118 Ma and since it has been affected by the movement in the shear zone, that age must represent the *oldest* possible age for movement along the fault. A few miles to the east (see Stop 1-4) lies the large La Posta pluton which will be discussed in a later chapter. The fault zone terminates at the pluton boundary (Stop 1-4) and must therefore be older than the pluton's emplacement age of 94 Ma.

STOP 1-4

END OF THE CLM FAULT ZONE ALONG KITCHEN CREEK ROAD. *Return to I-8 east and proceed southeast to the Kitchen Creek/Cameron Station exit (Figure 2.3). Turn north (left) along the paved Kitchen Creek Road. Drive 1.9 miles to a gravel pullout on the right (east) side of the road opposite some large, moderately fresh outcrops. These outcrops have been mapped as part of the same unit seen at Stop 1-3, and, although they do not have a zircon age, are a bit more accessible than the first stop. Again, this rock is a gneissic granodiorite with abundant biotite. The flakes of dark biotite, on close examination, are aggregates of smaller crystals strung out by movement in the fault zone. The foliation is similar to that of the gneissic rock at Stop 1-3 and is oriented to the northeast with a near vertical dip.*

The rolling valley floor east of the road at this stop is underlain by the huge 94 Ma La Posta pluton at whose contact the shear zone terminates, thus giving us the minimum age for fault movement. These younger rocks can be examined by backtracking about 0.4 miles towards I-8 to a paved left (east) turn that drops into an old concrete helipad. Park and walk over to several large bouldery outcrops adjacent to the south side of the helipad. These rocks belong to the western edge of the La Posta pluton (see Chapter 6, Stop 6-1) and are quite different in character from the rocks we just left. Notice that they are much lighter in color and have a "salt and pepper" character, i.e., they contain dark colored minerals (biotite and hornblende) set into a matrix of whitish feldspar and light gray quartz. Some of these dark minerals are rectangular in shape and roughly parallel to a few six-inch long scattered elongate dark inclusions. These dark inclusions (see Chapter 5, Stop 5-1) likely represent pieces of older rock picked up by the La Posta magma as it rose towards the surface. As the magma rose, these solid fragments and the few dark minerals in the magma at that point were rotated into parallelism as the magma sheared against the solid, older granodiorites. Compare a sample from this outcrop to those from the sheared granodiorite. Note that the biotite and hornblende are generally single crystals with rectangular or six-sided shapes as would be expected if they crystallized directly from a melt. Although the age of both outcrops is now well constrained by radiometric dating, geologists have long recognized that the strongly foliated rock must be the older of the two.

Field Excursion 2

Quartzite and Marbles at Dos Cabezas Siding

STOP 2-1

THE DOS CABEZAS QUARRIES. *Take I-8 east across the Peninsular Ranges and down the grade to the Ocotillo exit (S-2). Turn left (north) and drive to the stop sign at the far end of town. Continue from here on S-2 for 1.8 miles to the top of a long gentle upgrade. Just before the top of the hill, turn left onto an unmarked dirt road (Figure 2.6). This will take you to the marble quarries near the old Dos Cabezas railroad siding. The road is rough but easily passable with a truck or other high-clearance vehicle. Follow this road for 2.5 miles to a T-intersection just as you reach the railroad tracks. Turn right and follow this for 3.1 miles staying on the east side of the*

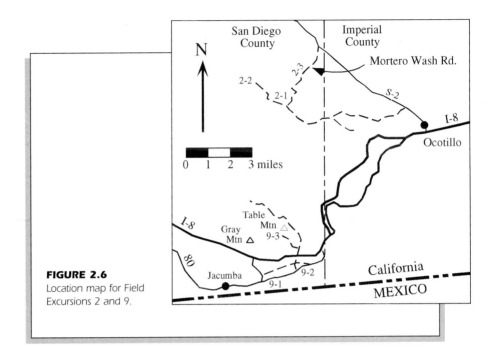

FIGURE 2.6

Location map for Field Excursions 2 and 9.

railroad track until you reach the old water tower at Dos Cabezas. From the flat area near the tower and old loading ramp, one of the old marble quarries can be seen about one-quarter mile to the northeast. A short walk or careful drive will bring you into the first of several small, abandoned open pits.

The whitish rocks exposed in the small quarry are fairly coarse-grained marbles consisting of white calcite plus a variety of other metamorphic minerals. The white fibrous mineral is wollastonite, a calcium silicate mineral formed by the reaction of calcite and small amounts of quartz in the original sedimentary rock during metamorphism, i.e., $CaCO_3 + SiO_2 \rightarrow CaSiO_3 + CO_2$. Since the parent rock contained much more calcite than quartz, the reaction consumed all the quartz leaving a calcite-wollastonite assemblage. The CO_2 released by the reaction eventually escaped into the atmosphere. Other metamorphic minerals found in this area include dark gray vesuvianite, orange garnet and pale blue-green diopside which also formed when the parent limestone reacted with bits of a clay mineral such as kaolinite. Above the quarry, low outcrops of very hard, dark colored rock are quartzites that consist almost entirely of dark colored quartz grains. These rocks represent relatively pure quartz sandstones that were interbedded with the adjacent limestones prior to metamorphism.

Marbles represent metamorphosed limestones, i.e., deposits of calcite that were precipitated either directly out of a warm, shallow ocean or by some type of biological activity, or both. Either way, these rocks represent a distinct change from the depositional environment of the schists and quartzites seen at Stop 1-1. They are found only in the eastern part of San Diego County as a discontinuous belt of ancient marble that extends southward into Baja California.

INDIAN HILL. *After leaving the marble quarries at Dos Cabezas, return to the flat area near the concrete loading platform along the railroad tracks. Follow the dirt road paralleling the tracks for about two miles to its end at a small dirt parking area. The track continues northward from this point for another mile or so before it makes a broad 180° turn for its southerly run through Jacumba Gorge. After that, it turns west to Campo and San Diego.*

Make the short walk to the railroad cut just above the parking area where more metasedimentary rocks can be examined. Small blocks of coarse-grained blue-gray marble are visible along the east side of the tracks but the large cuts on the western side expose several other rock types. The massive black material is amphibolite, a metamorphic rock composed of hornblende and plagioclase similar to the exposure on Sunrise Highway. It represents basaltic lavas intruded into the sedimentary layers before metamorphism. The amphibolite is surrounded by a deeply weathered gneissic rock that is thought to represent the remnants of a package of sedimentary rock that had been heated and metamorphosed to the point that melting began. The molten material drained off to form granitic dikes similar to the ones so prominent in the surrounding hills and left behind the biotite-rich gneissic rock as a refractory residue. The amphibolite, the coarse-grained marble, and a few pieces of quartzite scattered through the gneiss represent the bits and pieces of ancient basalt, limestone, and sandstone, respectively, that did not begin to melt, i.e., the most refractory rocks. As the melt was extracted and bled off into fractures, these rocks were rolled around like walnut pieces in a cake batter poured into a baking pan. Later, as temperatures waned, they froze into a series of isolated blocks within the gneiss.

At the southernmost end of the railroad cut is another massive but lighter-colored, metamorphosed sedimentary rock tilted steeply towards the north. Careful inspection of several thin layers high in the middle of this outcrop reveals a series of steep, tight folds that give an impression of intense swirling of solid but very plastic material. At a depth of about 14 kilometers (8.5 miles) and temperatures near 650°C, this now solid, brittle rock and those around it were as plastic as putty. Turn around and look at the hills to the east and north. The light-colored bouldery outcrops (Figure 2.7) represent the eastern edge of the small 89 Ma Indian Hill monzogranite, the youngest granitic rock in this area. The bulk of the pluton is behind you to the west and can be seen from the overlook at Sacatone Springs (Stop 7-2). The elongate, light-colored rock bodies are pegmatite dikes that are a bit older than the main body of granite. The dikes cut through the same gneissic unit you examined in the railroad cut and both are intruded by the bouldery monzogranite mass. This suggests that the relative age sequence at this stop, from oldest to youngest, goes from the gneiss (plus marble, quartzite and amphibolite) to the pegmatite dikes to the granite body itself. Thus, the dikes and the gneiss must be older than 89 Ma. Elsewhere in Southern California, similar pegmatitic dikes have been dated to between 95 and 100 Ma (see Chapter 5).

FIGURE 2.7 Bouldery outcrop of Indian Hill monzogranite. Two regular fracture directions within the outcrop appear to split the granite mass into separate blocks. A third direction, roughly parallel to the face of the outcrop, further breaks the rock into coherent blocks. This direction is best observed by climbing to the top of the outcrop. The large darker rock on the small peak to the left is a large inclusion of marble surrounded by migmatitic gneiss (see Field Excursion Stop 2-2).

STOP
2-3

MORTERO WASH. Retrace your route back to the Dos Cabezas area. From this point, you can return via the same road that brought you here or you can turn left (northwest) near the old loading area and return to County Route S-2 via Mortero Wash (Figure 2.6). This road is a bit rough in places and quite sandy once you enter the wash. However, it is mostly downhill and passable for the approximately four miles back to its intersection with S-2.

The first part of the drive takes you through a rolling terrane underlain by La Posta-type granitic rocks (see Chapter 5). Quartz, alkali feldspar, plagioclase, and biotite with minor sphene are the main minerals in this approximately 94 Ma igneous rock unit. As the road enters Mortero Wash, the outcrops that form the ridge above the north side of the road are coarse-grained white marbles. Marbles in Southern California commonly form topographic highs because the main erosional agent for calcite-rich rocks is dissolution by surface waters. In contrast, the igneous rocks have weathered through a combination of processes that weaken the boundaries between the different minerals causing the rock to literally disaggregate over time. The arid desert climate thus limits the weathering of calcite relative to the surrounding granitic and metamorphic rocks leaving the marbles as low ridges and hills standing above the granitic terrane. Near the irregular contact of these marbles with the La Posta-type granite are thin reaction zones (**skarns**) containing quartz, calcite, and pale-orange garnets. These developed as silica and aluminum from the granitic magma diffused into the solid marble. About a mile or so before the road intersects S-2, the wash cuts through dark hills underlain by black, vesicular volcanic rocks of Miocene age that are related to the rotational forces that disrupted Southern California beginning about 20 million years ago (see Chapter 6).

Late Jurassic— Early Cretaceous

The Santiago Peak Volcanics

 Magmatic Arc Processes

V olcanic arcs and their underlying plutonic roots are a result of complex interactions between lithospheric plates. These plates average about 100 kilometers in thickness and consist of a thin veneer (5–10 km) of basaltic crust that overlies approximately 90 kilometers of cool, rigid, and brittle mantle peridotite (see Appendix A), a very dense rock consisting of heavy minerals such as olivine and pyroxene. Beneath this lies the asthenosphere (Figure 3.1). It consists of similar solid peridotite that, due to the Earth's internal heat engine, is hotter, more plastic, and thus capable of flowing slowly under stress. The asthenosphere undergoes thermal convection that causes hot, plastic peridotite beneath the mid-ocean ridges to rise. This forces the cooler rock near the top of the asthenosphere to move laterally away from the ridge axis. The cooler rock will eventually sink back deeper into the mantle to reheat and complete the convective cycle. These circular motions in the upper asthenosphere drag on the base of the brittle, rigid lithospheric plates and carry them on their journeys around the globe.

Below the oceanic spreading ridges, upwelling asthenospheric mantle partially melts to create new basaltic oceanic crust that, along with the upper 100 kilometers or so of the brittle mantle lithosphere, is moved in conveyor belt-like fashion towards distant collision zones. Pressure, in all geologic systems, is equal to the weight of the overlying column of rock. As the plume of hot, solid mantle rises towards the ridge axis, total pressure on the plume decreases. Experimental studies have shown that if pressure is lowered, the temperature *at which the rock will begin to melt* likewise falls. When the new melting point lowers to reach the ambient temperatures in the rock, melting will begin. This phenomenon is similar to what backpackers and campers have all experienced. It takes longer to cook a hard-boiled egg high atop mountain ranges than it does at sea level. This is because the pressure on the water (weight of the column of overlying air) is less and the boiling temperature of the water drops accordingly. Lower temperatures mean longer cooking time. Lower pressures on rock systems means lower melting temperatures.

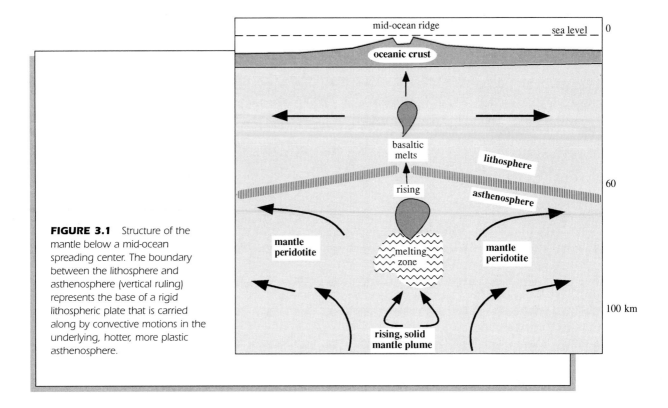

FIGURE 3.1 Structure of the mantle below a mid-ocean spreading center. The boundary between the lithosphere and asthenosphere (vertical ruling) represents the base of a rigid lithospheric plate that is carried along by convective motions in the underlying, hotter, more plastic asthenosphere.

At the spreading ridges, a process takes place that has great influence over the phenomena that occur at subduction zones. Lying under several kilometers of ocean water, cold seawater penetrates through fractures and faults towards the underlying basaltic magma chambers (Figure 3.2). These marine waters become heated to as much as 400°C and rise back through fissures in the newly created volcanic rock. As these saline geothermal waters pass through the new oceanic crust, they partially destroy the original mineralogy in the basaltic rocks and convert it to hydrous minerals such as clay and serpentine. The geothermal solutions quickly rise to the ocean floor where they cool by venting as submarine geysers into the cold ocean waters. As spreading continues, the plates and the newly created, but altered, oceanic crust move away from the ridge axis. Within a few kilometers, access to the heat of the magma chambers is denied, fractures become clogged with precipitated material from now cooler solutions, and the alteration process effectively ceases. The altered crust is eventually capped with thin layers of sediment and continues on its journey towards some distant collision zone at rates of only a few centimeters per year. So vigorous is the hydrothermal system at the ridge axis that the *entire volume of ocean water* may circulate through newly created oceanic crust in just a few million years. This results in an altered basaltic crust in which modest amounts of seawater are stored not in fractures as free water but in the structure of the newly formed minerals. This stored water, once released in a subduction zone, controls not only the type of magma that forms but the potential for catastrophic eruptions in the overlying volcanic arc.

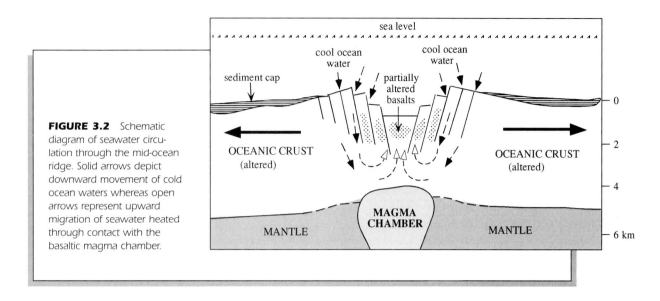

FIGURE 3.2 *Schematic diagram of seawater circulation through the mid-ocean ridge. Solid arrows depict downward movement of cold ocean waters whereas open arrows represent upward migration of seawater heated through contact with the basaltic magma chamber.*

When lithospheric plates collide (Figure 3.3), the downward motion of the subducted plate is a combination of its forward horizontal motion combined with sinking of the colder, denser plate into the warmer, less dense asthenosphere. The angle that the downthrust plate makes with the surface is a function of the rate at which the plate is being pushed, i.e., the rate of spreading at the ridge crest. The faster the spreading, the shallower the angle of subduction. Once subduction has begun, any variation in the spreading rate will change the angle of subduction and thus the physical conditions under which magmas are produced in arc systems.

As the colder lithospheric plate descends further and further into the asthenosphere, the upper portions, i.e., the hydrothermally altered basaltic crust encounters hotter and hotter environments and corresponding increases in pressure. As pressure and temperature increase during subduction, the rocks begin to undergo a progressive series of changes in mineralogy. At each step, new minerals appear that contain smaller and smaller amounts of water and other volatiles in their structures. The alteration minerals created at the mid-ocean ridge geothermal zone are metamorphosed to less hydrous counterparts. Clays are eventually converted to feldspar and serpentine to pyroxene. Most of the volatiles, mainly water, are released during this metamorphism and can either remain as a free volatile within the oceanic crust or leak upwards into the overlying mantle wedge.

Volatiles such as water play a critical role in the development of volcanic arcs. If water exists as a free phase rather than being tied up structurally in hydrous minerals, it acts as a fluxing agent and drastically lowers the melting temperature of the enclosing rock. Dry mantle peridotite, for example, begins to undergo partial melting at around 1500°C. If water is somehow added to that rock, melting begins at 1200°C and will continue until it has dissolved into the newly created melt. Thus, the melting temperature and degree of partial melting of any source rock are dependent upon the amount of free water

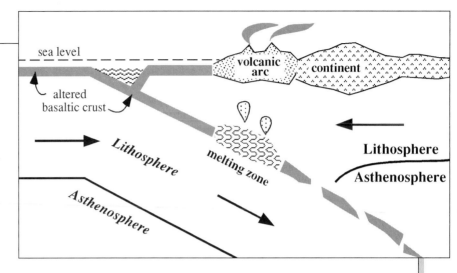

FIGURE 3.3 *Schematic representation of a subduction zone that creates a volcanic arc adjacent to a continental margin. As one lithospheric plate is thrust underneath the other, its cap of altered basalt is forced into hotter and hotter portions of the mantle. Sea water trapped within the alteration minerals is released through metamorphism and lowers the melting temperature of the basalt and overlying mantle wedge. The depth of this melting zone and the position of the volcanic arc relative to the continent are thus dependent upon the angle of subduction. A steeper subduction angle creates a volcanic arc closer to the trench whereas a shallower angle causes the arc to form inland on the continent.*

available. It also means that the melts produced by this process will contain large amounts of dissolved water as they rise towards the surface.

In order to create the wide range of rock types characteristic of magmatic arcs, several lithospheric sources must be available for melting. The basaltic component of the island arc suite can only be produced by partial melting of the mantle wedge immediately over the subduction plane. Although this material is peridotite and similar to the mantle materials melted below mid-ocean ridge crests, the basaltic rocks found in volcanic arcs are *slightly* different in composition, suggesting that the conditions that create arc basalts differ from those at the mid-ocean ridge spreading centers.

Arc basalts erupt in a more spectacular and energetic fashion than ridge or "hot spot" basalts. This eruptive style can be directly related to the increased amount of water dissolved in arc-related melts. Melts rise buoyantly towards the surface because they are less dense than the surrounding rocks. Their ability to retain the dissolved water is a function of pressure or depth below the surface. At shallower levels, the melt cannot keep the water in solution and it begins to separate into small "bubbles" within the silicate melt. At near-surface pressures, melts are incapable of carrying *any* dissolved volatiles and the melt begins to "effervesce" rapidly. The "bubbles" coalesce and rise buoyantly through the magma column pushing the melt ahead towards the surface vent. This process gives rise to spectacular eruptions with "fire fountains" and great bursts of explosive activity, all of which is due to the release and expansion of the water dissolved within the melt. Several kilometers below the volcanic edifice, the water-bearing basaltic melt may congeal to form coarse-grained plutonic rock (gabbro) but the water, which cannot escape at that depth, remains in the melt. Eventually, it is incorporated into the rock in the form of water-bearing (hydrous) minerals such as amphibole. It is this mineralogical characteristic, the presence of amphibole, that distinguishes arc-related gabbroic rocks from those that form in the magma chambers beneath spreading ridges, a distinction that will become important in the following chapter.

The bulk of the island arc suite, however, falls in the andesite-dacite range. Although still hotly debated, most petrologists consider this suite to be the result of melting of downwarped water-rich oceanic crust that has been dehydrated through metamorphism. The released water lowers the melting temperature of reconstituted oceanic crust which partially melts to create the andesite-dacite suite that is found only within volcanic arcs. Again, with greater amounts of water dissolved in these relatively viscous melts, its release at volcanic levels creates the spectacular and devastating eruptions so characteristic of volcanic arcs. From Tamboro in 1815, to Krakatoa in 1883, Mount Saint Helens in 1980, and Mount Pinatubo in 1991, these volcanic systems have left their mark on human history.

△ The Santiago Peak Volcanics

Westward from the Late Triassic-Jurassic Cuyamaca-Laguna Mountain arc lies another ancient magmatic arc system, the Santiago Peak Volcanics (SPV). It runs southward from Riverside County to the Agua Blanca fault in northern Baja California in a belt up to 20 km in width (Figure 2.1). This system, however, is mainly composed of volcanic flows and **breccias** (rocks composed of broken fragments of other rocks) along with subordinate amounts of sedimentary materials and minor shallow-level intrusions. It is, in effect, the remnants of a volcanic arc that has not undergone the deep erosion that, for example, removed the volcanic components of the older CLM arc. These rocks were named after a series of exposures at Santiago Peak in the Santa Ana Mountains of Riverside County but are found in abundance in San Diego County at such locations as Otay Mesa, Del Cerro, Black Mountain, and Rancho Santa Fe. They consist of a wide variety of rock types that range from basalt through andesite and dacite to rhyolite (see Appendix A). The bulk of this rock suite, however, falls within the andesite-dacite compositional range with basalt and rhyolite making up only a small proportion each of the total rock volume. This wide range of magma compositions is typical of most volcanic arcs and attests to the complexity of processes and source rocks found within a subduction setting.

The SPV represent an ancient island-arc system generated by the collision of two lithospheric plates. Rocks exposed from the Santa Ana Mountains to the Agua Blanca fault south of Ensenada in Baja California exhibit textural and compositional changes from predominantly basaltic flows in the north to a breccia-dominated andesite-dacite suite in the south. In Riverside County, some of the SPV is interbedded with the Bedford Canyon Formation, a series of latest Jurassic sedimentary rocks, and suggests that the volcanic system is no older than Jurassic. This older portion of the SPV has been assigned a Late Jurassic (~ 145 Ma) age based on fossils found in the associated sedimentary rocks. Recent zircon U-Pb ages from all other areas, however, range from 128 Ma to 116 Ma and indicate that the main portion of the SPV is somewhat younger. In addition, these magmas do not appear to have interacted, as did those of the CLM arc, with the ancient rocks of the North American craton and must have been intruded westward of that terrane through oceanic lithosphere (Figure 3.3).

Volcanic arc rocks around the world are dominated by explosive eruptions and fragmental textures. The SPV is no exception. Fragmental rocks such as tuffs and tuff-breccias (see Appendix A) are the most common rock types observed, with volcanic flows making up less than 20% of the total rock volume. The original SPV rocks contain visible crystals of plagioclase, quartz, and pyroxene in a dark, fine-grained matrix. They have been partially converted to alteration minerals such as epidote, chlorite, and micas. This alteration mineral assemblage is probably due to circulating hydrothermal solutions that are known to create sulfide-rich metal deposits within volcanic arcs. Although San Diego County is not known for such deposits, scattered remnants of these ore-forming systems still exist in areas near Olivenhain, El Cajon, and Otay Mesa and may represent fragments of larger metal deposits stripped away by erosion.

In the Black Mountain area some 30 km north of downtown San Diego, more than 3 km of total stratigraphic thickness of SPV rocks have been mapped with evidence for both submarine and subaerial eruptions. It seems, therefore, that the portions of the SPV observed within San Diego County most likely represent a volcanic arc emerging from moderate ocean depths well offshore from the main landmass of the North American continent. Spectacular eruptions may have carried volcanic ash large distances eastward into the shallow continent-fringing seas and over the older (Triassic-Jurassic) CLM arc which by now had ceased eruptive activity and was being dissected by erosion. Metallic ore deposits similar to those found along the northern portions of the Japan magmatic arc likely formed as circulating marine waters, heated by volcanic activity, redistributed the trace amounts of metals found within the volcanic pile. The materials on which San Diego County topography would later be carved were beginning to come together.

Field Excursion 3

Santiago Peak Volcanics

THE DEL CERRO BLOCK. From I-8, take the College Avenue exit north to Del Cerro Boulevard. Turn right and proceed two short blocks to Madra Avenue. Turn left and head up and over Del Cerro about 1.3 miles to the entrance to Lake Murray Park (Thomas Brothers Map 1250, E-6). Lake Murray is at the southern end of the Mission Trails Regional Park, a 5,800-acre parcel of hiking trails and other recreational facilities created in 1974. Park as close as you can to the east-facing slopes bordering the lake. Walk to the large outcrops above the jogging road that circles the lake.

These outcrops are typical of the SPV in the Del Cerro block and in the general San Diego area. On close examination, the rocks are dark and fine-grained but hold a wide variety of angular pieces of other lighter colored volcanic (and sparse plutonic) rocks (Figure 3.4). The volcanic fragments are characterized by a porphyritic texture, i.e., two populations of grain sizes. One is so small that individual grains are not recognizable (the dark matrix) and the other large enough so that individual, well formed crystals of feldspar,

FIGURE 3.4 Outcrop of Santiago Peak Volcanics along the west side of Lake Murray. The rock consists of a dark matrix of glass and finely ground rock material plus large fragments of other porphyritic volcanic and sparse plutonic rocks. It formed as volcanic rock and ash were ejected and mixed together along the flanks of a growing composite volcano.

hornblende, or even quartz can be identified. This texture is characteristic of magmas that have undergone two stages of cooling, the earliest of which was slow enough to allow crystals to begin to form and grow before the magma was quenched by eruption to the surface.

These tuff-breccias (Appendix A) are common in volcanic arcs and are the result of the explosive eruptions that characterize these systems. Fine ash, bits of still molten lava, pieces of the volcanic edifice itself, and solid fragments of the earlier pulses of volcanism are ejected during an eruption and settle back to form a blanket of broken fragments. Some of this material deposited high on the slopes of the volcano becomes saturated with water from accompanying rainstorms and breaks loose to roll downslope, blending ash and many different pieces of rock into the outcrops that we see today. The volcanic landforms are long since gone. Only the most resistant rocks remain and they bear testimony to the volcanic processes that created this part of San Diego County's geologic past.

STOP 3-2

COWLES MOUNTAIN. The highest point in the City of San Diego at 1,591 feet above sea level, Cowles Mountain was named for one of San Diego's early ranching pioneers, George Cowles, who settled in El Cajon valley in 1873. A rewarding hike of 1.5 miles from the trailhead at the corner of Navajo Road and Golfcrest Drive (Thomas Brothers Map 1250, F-4) to the top of the mountain will give the visitor a spectacular 360° view of San Diego.

The change in elevation for this hike is about 1,000 feet. Bikes are not allowed on this trail.

*T*he lower slopes of Cowles Mountain are covered with colluvium, i.e., rock debris from the higher elevations that has moved downhill forming a blanket of loose material. The first outcrops are encountered just past the 1/4-mile marker and are quite different from those seen at the previous stop. The outcrops along the trail are fine- to medium-grained plutonic rocks that likely represent the last magma pulse to cool within the volcanic edifice. Good outcrops occur just before the 1/2-mile marker at the first shoulder of the mountain along the trail. Here you can see that the average grain size is roughly 2 mm. With a hand lens or small magnifying glass, you can see equal-sized grains of black hornblende, gray quartz, whitish plagioclase and vaguely pink alkali feldspar. The hornblende and feldspar grains tend to weather out of the rock leaving small, rectangular voids. The outcrops also contain small (2–5 cm) dark fragments of older, very fine-grained volcanic rocks.

Continue up the trail toward the summit. About 1/4 mile ahead and just before you reach the 3/4-mile marker, look up above the trail to see a series of columnar outcrops with smooth, almost sculpted forms. At the 3/4-mile marker, look left across a small ravine at a series of outcrops with a very distinctive fracture pattern. Continue ahead past the cutoff to Barkers Way to a sharp bend in the trail (just below the 1-mile marker). The prominent outcrop at the bend is a good example of the fracture patterns in the rock unit. The most prominent fracture set is vertical and points uphill towards the summit but three other fracture sets can also be observed. The second is also vertical. It is essentially parallel to the mountain front at this point and at a right angles to the first. A third, less apparent, set is almost horizontal while a few examples of a fourth set appear as about 45° to the first. The combination of the first three fracture directions gives rise to the column-like appearance to the outcrops along this portion of the trail. Part of the sculpted or smooth appearance to some of these outcrops may to be due to circulation of heated groundwater through the fracture system. As discussed in the preceding section on volcanic arc processes, surface waters penetrate the heated interior of the volcanic edifice through these fractures and begin to rise back towards the surface. As they do, they dissolve the more soluble elements out of the rocks and redeposit them in the cooler (higher) portions of the fracture system. This includes relatively soluble elements such as potassium and silica as well as metallic elements such as copper, lead, gold, and silver. Many of these ore-forming systems around the world have a cap zone in which the original minerals in the rock have been replaced by very fine-grained silica (quartz). Close examination of some of the clean fracture surfaces reveals, to a practiced eye, a vitreous or glassy sheen reminiscent of the luster of quartz. Silica alteration? During the early 1990s, core drilling as part of a feasibility study for a water diversion tunnel underneath the western side of Cowles Mountain revealed veinlets of calcite and other minerals that precipitate from geothermal solutions. If so, does this mean that there are ore bodies beneath Cowles Mountain? Perhaps! It is likely, though, that some fine-grained silica alteration has occurred along the fractures. This would increase the resistance of the rock to erosion and help to produce the sculpted appearance of the outcrops along the trail.

At the summit area, about 1/2 mile ahead, look again at the rocks. Here, the grain size is distinctly larger, perhaps 3–4 mm on average but it is the same rock. This suggests that we have also moved laterally into the more slowly cooled core of the Cowles Mountain complex. Imagine, if you will, that the rocks were formed in the throat of a large volcanic edifice and that they represent the last surge of magma into the volcanic conduit. It crystallized in place and cooled relatively slowly, losing heat evenly to the surrounding volcanic cover. The fracture patterns were produced by the slow cooling and regular shrinkage of the rock mass. The dark inclusions represent pieces of the older volcanic pile through which the enclosing magmatic pulse rose. Look down towards Lake Murray and visualize the rocks seen at the first stop (3-1). If you missed that stop, just go to the old marker at the summit. The rocks that make up the marker are identical in form and may represent parts of the vented material that built up the volcanic edifice. This material has been stripped away by erosion, leaving the more resistant, coarser-grained rocks as a topographic high while remnants of the flanking flows and breccias are now found at lower elevations surrounding the peak (Figure 3.5).

The summit provides a marvelous 360° view of the greater San Diego area.

Go over to the western-view informational sign and pick out Black Mountain and North Fortuna Mountain. These peaks are also underlain by similar rocks. Are they the erosional remnants of other volcanoes or were they part of a single volcanic structure centered around Cowles Mountain? At the eastern-view informational sign, you can pick out Otay, San Ysidro, and San Miguel Mountains, all built out of similar volcanic material. From this vantage point, you can also get a good view of the high peaks in the Peninsular Ranges batholith. Cuyamaca and McGinty Peaks, and Viejas and Los Pinos Mountains are underlain by resistant gabbroic rocks while Lyons Peak is composed of

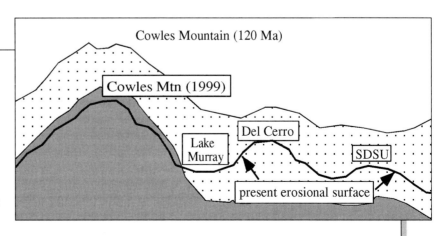

FIGURE 3.5 A schematic cross section showing a possible relationship between the tuff-breccias in the Del Cerro-San Diego State University area (dot pattern) and the intrusive rocks at Cowles Mountain (shaded). The heavy dark line represents the present erosional surface. The volume of rock above that line has been stripped away by erosion.

monzogranite (see Chapter 6). Both rock types are resistant to the forces of weathering and erosion in this environment and weather into topographic relief as the less resistant rocks around them are stripped away. Can you imagine the volcanic debris of the SPV stretching out to cover all of these rocks? Are those peaks also core zones to now eroded volcanoes that produced extensive volcanic covers? A zircon U-Pb age of 122 Ma from Cowles Mountain suggests that both systems overlapped in time (see Chapter 6).

BLACK MOUNTAIN. Exposures of banded andesite flows occur along Black Mountain Road just north of Black Mountain Park (Thomas Brothers Map 1149, E-7). The easiest access is I-15 to Rancho Bernardo Road west which turns into Black Mountain Road after about one mile. Continue west another 1.5 miles to a sharp left turn and follow the unpaved road south. At about 0.9 miles from the turn, low outcrops of the Eocene Poway Conglomerate occur on both sides of the road. These sedimentary rocks were deposited on top of the Jurassic SPV by a large Eocene river system that cut through the weathered Cretaceous igneous rocks to the west. The rounded porphyritic volcanic clasts within the outcrop were brought into this area from north-central Mexico prior to the opening of the Gulf of California. The sandy matrix is what remains of the deeply weathered granitic rocks through which the river ran.

Four-tenths of a mile ahead is the north end of a long outcrop of flow banded andesites. The banding is visible as thin white to gray layers that pinch and swell along strike with thicknesses ranging from a few millimeters to several centimeters. These color variations are due to microscopic bits of quartz, epidote, and pyrite that may have been created during either metamorphism or hydrothermal alteration of this rock unit. The rock also contains tiny microphenocrysts of plagioclase feldspar set into a very fine-grained volcanic matrix consisting of feldspar and quartz. At this point, the outcrop contains two subparallel fault planes about one meter apart dipping about 30° to the south. The cleavage in the banded andesite above and below the two shear planes dips gently northward at about 25°. In between, the cleavage is consistent with a clockwise sense of rotation that suggests these rocks were being compressed (pushed together), as would be expected in a volcanic environment that marked the collision of two tectonic plates.

CHAPTER 4

THE CRETACEOUS

 The Peninsular Ranges Batholith

The central and eastern portions of San Diego County contain a series of small north-northwest trending mountain ranges that are part of the larger Peninsular Ranges geomorphic province (Figure 1.2). Familiar names such as the Laguna, In-Ko-Pah, and Jacumba mountains are found on all county maps. The mountain ranges are part of the Peninsular Ranges Batholith (PRB), a complex mixture of igneous and metamorphic rocks that extends southward from Riverside, California, to about the 28th parallel in Baja California. The batholith averages nearly 100 km in width and has a north-south length of approximately 1,000 km. It is composed of discrete bodies of coarse-grained igneous rock (**plutons**) that vary from a few kilometers to more than 40 kilometers in diameter as well as metamorphosed remnants of the Jurassic and pre-Jurassic rocks described earlier. The igneous rocks within the PRB reflect a complex and nearly continuous history of igneous activity that began near 125 Ma and lasted until about 90 Ma. That these rocks are related to subduction processes is not disputed. Similar rocks are found in the Salinian block of west-central California and in the great Sierra Nevada Batholith (Figure 1.2). They indicate that during Mesozoic time, an extensive magmatic arc formed the western edge of the North American continent. This arc was part of a larger system that outlined much of the eastern Pacific Basin. It extended as far north as Alaska and southward along the western edge of South America to Tierra del Fuego at the southern tip of Argentina. The Mesozoic was a time of extensive volcanic activity during which plate motions, driven by changes at mid-ocean spreading centers, were rapid and variable. Thus, the Mesozoic history of greater San Diego County reflects that of the entire western Pacific and was irrevocably tied to fluctuations in spreading rates at mid-ocean ridges hundreds of kilometers distant.

The PRB contains hundreds of individual plutons, and countless dikes and small igneous bodies that were intruded into preexisting crustal rocks such as the Julian Schist and the gneissic granitic rocks in the Cuyamaca-Laguna Mountain (CLM) belt. It has an elongate, somewhat curved shape similar to that of a volcanic arc created by the collision of two lithospheric plates and is considered to be the root zone of such a crustal structure. In fact, the PRB consists of the plutonic roots of two adjacent parallel arcs that formed sequen-

33

tially in response to changes in plate interactions. The boundary between these two systems occurs just east of the Laguna Mountains and trends northwest-southeast (Figure 4.1). It divides the batholith (and the county) into an older western and a younger eastern zone that have their own distinct rock types and structures. The geologic history of each zone is thus the story of a single period of plate collisions that created the separate pieces of the batholith. We view this history by examining the rock types in each zone and then reconstructing the processes that likely created them.

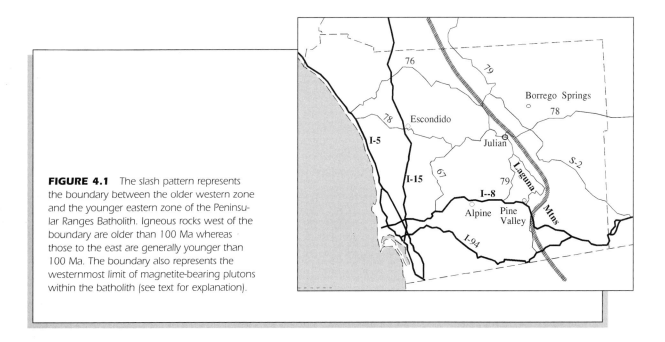

FIGURE 4.1 The slash pattern represents the boundary between the older western zone and the younger eastern zone of the Peninsular Ranges Batholith. Igneous rocks west of the boundary are older than 100 Ma whereas those to the east are generally younger than 100 Ma. The boundary also represents the westernmost limit of magnetite-bearing plutons within the batholith (see text for explanation).

△ The Western Arc: Diversity and Dynamics

Three main plutonic rock types—gabbro, tonalite, and monzogranite (see Appendix A)—occur in the western zone of the batholith and form separate plutonic bodies that can be recognized in the field by their distinctive morphology. The earliest magmatic activity in this zone of the batholith is represented by rocks of gabbroic composition. They have been eroded into topographic prominence to form landmarks such as Cuyamaca Peak (Figure 4.2), Tecate Peak, Viejas Mountain, and Los Pinos Mountain, and have sparse scattered dark-colored outcrops with dark gray to red soils. The tonalitic plutons underlie relatively low, rolling terranes with scattered dark gray outcrops and grayish soils. The area immediately around the town of Alpine, above which the gabbroic Viejas Mountain rises in prominence, is a good example of the different weathering styles of these two rock types (Figure 4.3). Roadcuts or large outcrops of tonalite can also be recognized because they commonly contain dark, crudely aligned, ellipsoidal features up to two feet in length

FIGURE 4.2 Aerial view looking west at the gabbroic Cuyamaca (6,512'), Middle (5,883') and North (6,000') Peaks. The light-colored rock mass in front of Cuyamaca Peak is the monzogranite of Stonewall Peak (5,730').

FIGURE 4.3 The dark gabbroic outcrops of Viejas Mountain tower above the bouldery, light-colored monzogranite and low-lying gray tonalite.

called xenoliths (Figure 4.4). The xenoliths are pieces of older, finer-grained igneous rock that formed along the quickly cooled margins of the pluton. They were broken off and swept into later pulses of the magma as the chamber began to fill.

The monzogranite plutons also form prominent ridges and peaks such as Lawson Peak near Jamul, Stonewall Peak in Cuyamaca Rancho State Park, and Woodson Mountain near Ramona. They are characterized by light-colored, bouldery outcrops that appear to have been piled up like marbles by some mythical creature. Nothing can be further from the truth. The different topographic expressions of these three rock types can be traced back to somewhere between the Late Cretaceous and Early Eocene (see inside front cover) when the climate was much wetter than today. Warmer temperatures, elevated con-

FIGURE 4.4 *Swarm of dark inclusions in tonalite. Their shape and preferred orientation indicate that the inclusions were solid rock fragments incorporated into the rising tonalite magma.*

centrations of atmospheric carbon dioxide, and abundant decaying vegetation combined with the rainfall to chemically weather all the near-surface rocks. Feldspars were partially converted to clay minerals while biotite and amphiboles altered to a mixture of clays, silica, and soluble materials. The degree of chemical weathering was, in good measure, controlled by the ability of these solutions to penetrate into the solid rock. This was, in turn, controlled by the degree of fracturing in the three rock systems. The gabbros and monzogranites appear to be less fractured and to have undergone less intense chemical weathering than the tonalites so that erosion by running water would preferentially remove the weathered tonalite debris leaving the other two rock types as topographic highs. Many of the monzogranite plutons, however, developed three-dimensional, intersecting fracture patterns. Weakly acidic solutions (H_2CO_3), a result of rainwater combining with carbon dioxide in the atmosphere, sank into the fractures and destroyed the corners and edges of the fracture blocks first to leave a rounded corestone of solid rock surrounded by the decomposed and more easily eroded products of chemical weathering. As the decomposed material was stripped away, large rounded boulders were left behind to form what appears to be boulder-strewn hills. Mount Woodson, near Ramona, is an excellent example of this process (Figure 4.5).

FIGURE 4.5 *View of Mt. Woodson from Hwy. 67. The bouldery outcrops are a result of deep chemical weathering along a three-dimensional fracture array with erosion stripping out the loose material between the less weathered cores of the fracture blocks (see text).*

Radiometric ages on the gabbroic rocks have been difficult to measure so that their crystallization ages are usually given in relative terms. Most investigators have agreed that the individual gabbroic plutons are older than the granitic rocks (tonalite and monzogranite) immediately adjacent to them. These determinations, however, are based on interpretations of field relationships which, in many cases, are either ambiguous or hidden beneath brush, pavement, or soil cover. The fact that these rocks weather to form prominent peaks also means that their lower slopes are covered by debris from modest-sized landslides which obscure their contacts with adjacent plutons. By plotting the distribution of rock types in the form of a geologic map (Figure 4.6), relative age determinations can be made in one of several ways:

1. Dikes of one rock type cutting through a second indicate that the dike rock is the younger.
2. Pieces of one rock type (xenoliths) in a second argue that the enclosing rock is younger.
3. Layering, banding, or any type of directional feature in one rock that is terminated at the contact with a second rock indicates that the second rock is the younger.

Using such observed relationships, the gabbroic rocks within the western zone of the PRB have been shown to be consistently **older** than the adjacent granitic rocks. This statement, however, must be taken with some caution in that it does not necessarily mean that all gabbroic plutons are older than all the granitic plutons within the batholith, just those in their immediate vicinity. Similar age relationships are reported between the gabbroic and granitic rocks in the Sierra Nevada batholith, the Coast Range batholith of British Columbia and many other batholithic terranes around the Pacific Basin.

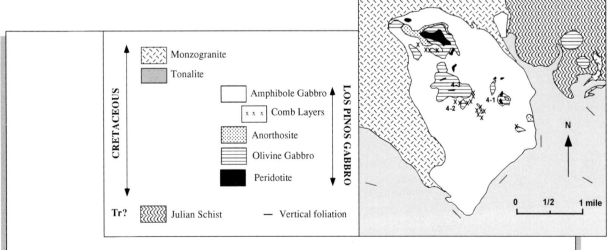

FIGURE 4.6 Geologic map of the Los Pinos pluton with Stops 4-1, 4-2, 4-3 approximated. The monzogranite (slash pattern) crosscuts Julian Schist, tonalite, and Los Pinos gabbro and is the youngest of the mapped rock units. The tonalite has a moderately well developed northwest trending foliation that wraps around the southern end of the gabbro indicating that it intruded into and around the older gabbroic pluton. (From **Canadian Journal of Earth Science**, Vol. 13, No. 9 by Michael Walawender. Reprinted by permission of National Research Council of Canada.)

Gabbroic Rocks

*G*abbroic plutons are found only in the western zone of the batholith (Figure 2.1) and underlie about 15% of the land surface. They are typically small bodies with exposed areas less than 10 kms[2] that can be easily identified through their topographic prominence, lack of bouldery outcrops, and dark reddish soil. Cuyamaca Peak (Figure 4.2), Tecate Mountain, Viejas Mountain (Figure 4.3), and Los Pinos Mountain (Figure 4.7) are all gabbroic peaks. Scattered dark outcrops occur only higher up on the slopes with small amphitheater-like depressions in mid-flank attesting to landslide-controlled mass wasting as one of the main erosional mechanisms leading to their destruction. These features are apparent along the south-facing slopes of Viejas Mountain (Figure 4.3) located on the north side of I-8 about 15 miles east of San Diego.

FIGURE 4.7 *Aerial view of the Los Pinos gabbro pluton (dark mass, right center). The lighter-colored outcrops are monzogranites with Corte Madera peak in the upper center. The low area in bottom center is underlain by less erosionally resistant tonalite.*

Gabbroic rocks form in essentially two types of tectonic settings: magma chambers beneath spreading centers and island arcs above a subduction zone. Those that form in subduction settings are mineralogically distinct from those that form beneath spreading centers. Although olivine, pyroxene, and plagioclase are major constituents of these rocks in both settings, subduction-generated gabbroic rocks also contain significant amounts of amphibole and thus stand apart. Amphibole is a hydrous mineral whose presence indicates that the parent melt must have contained several weight percent water. As described in an earlier section, island arc basalts and their plutonic equivalent, gabbro, are generated where volatiles, released through the metamorphism of altered oceanic crust in a subduction zone, filter into the overlying mantle wedge (Figure 3.3). The volatiles, mainly water, lower the melting temperature of overlying mantle peridotite to the point where melting begins and then dissolve into the newly generated melt. These early melts must therefore have a relatively high water content and must eventually crystallize a hydrous mineral such as amphibole.

Within the gabbroic plutons, two distinct rock suites are recognized. The first is more mineralogically diverse and contains, in the order that they began to crystallize from the basaltic melt, spinel, plagioclase, olivine, two types of pyroxenes, magnetite, and amphibole. Gravity-controlled settling of the early formed minerals into small pockets within the magma chamber gave rise to a variety of rock types that range from peridotite through gabbro to anorthosite (see Appendix A). Olivine, pyroxene, and spinel are more dense than plagioclase and sank towards the bottom of the magma chambers to form peridotite. Plagioclase followed suit but, because of its lower density, did not sink as fast and accumulated above the peridotite in an undisturbed magma chamber to form light-colored anorthosite. Figure 4.6 shows a series of small dike-like bodies of peridotite and anorthosite within the Los Pinos pluton. These could only have formed if the magma chambers were repeatedly recharged with fresh parental melt from the source region. As the chamber swells during this filling event, the early crystal aggregates, peridotite and anorthosite mushes, are pushed and squeezed into fractures where they finish crystallizing into the observed rock types.

The second suite consists of only amphibole and plagioclase. These rocks are typically very coarse grained, a feature generally attributed to an abundance of water in the parent melt (see Stop 4-2). They formed from a basaltic melt that was chemically distinct from the one that crystallized into the olivine- and pyroxene-bearing rocks *only* in the sense that it contained enough dissolved water so that anhydrous minerals such as olivine and pyroxene could not form. These rocks vary from fine grained to pegmatitic with abundant crosscutting relationships (Figure 4.8) and again suggest that the gabbroic plutons were created by multiple injections of melt. Of the two rock systems, the amphibole-plagioclase gabbros are the least abundant at the present erosional level but, as will be discussed later, may be the most important.

The gabbroic rocks have been subjected to the same amount of erosion (about 9 or 10 km) as the western quartzite-schist sequence of the Julian Schist and are far removed vertically from their volcanic cover. However, some structures within these plutons indicate that the magma system did, in fact, vent

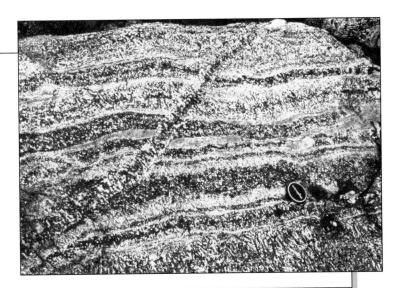

FIGURE 4.8 Comb layers consisting of elongate black amphibole crystals oriented perpendicular to the layering. (See Stop 4-2 for discussion.) Note the comb-layered dike crosscutting the main set of layers.

to the surface. The gabbroic bodies were created by multiple injections of very similar melt that repeatedly crystallized the same sequence of minerals, a pattern identical to that observed in basaltic volcanoes around the world. In addition, thick layers of very coarse-grained gabbro alternate with thin layers of fine-grained gabbro at Los Pinos Mountain (Stop 4-2). The latter material appears to have formed during periods of reduced pressure on the melt that could be a product of surface eruptions as much as nine kilometers above the present summit.

Zircon, the mineral that is the primary source for the uranium used in determining radiometric ages for plutonic rocks, is scarce or absent in gabbroic rocks. As of this writing, only two preliminary ages of 109 and 122 Ma have been reported from gabbroic rocks in San Diego County. Other radiometric techniques such as rubidium/strontium or neodymium/samarium have proved equally frustrating because of the very low abundances of those elements in gabbroic rocks. In effect, the emplacement ages of this rock system can only be estimated from the emplacement ages of the spatially associated granitic rocks. Although the gabbroic rocks appear to represent the oldest plutonic assemblage in the PRB, the timing of the onset of magmatism is uncertain and is a crucial point as we begin to look in detail at the evolution of an ancient island arc system.

Field Excursion 4

Los Pinos Mountain

MAGMA CHAMBER. *Take the Morena exit from I-8 and turn right (south) on Buckman Springs Road (Figure 4.9). Proceed south across Old Highway 80 approximately 2.3 miles to the paved Morena-Stokes Valley road. Turn right (east) and follow the road past the honor camp where the pavement ends. Continue on the gravel road along the east side of Morena Reservoir for about 3 miles to the Four Corners ORV area. Turn right on the gravel road toward the Los Pinos summit and proceed about 1 mile to a pullout at the first sharp left turn in the road. The bowl-shaped area to the east below the pullout exposes a wide variety of gabbroic rocks, including dark-colored outcrops of peridotite and olivine gabbro just below the road, and obvious white outcrops of anorthosite and anorthositic gabbro about 150 meters further down the slope.*

These outcrops were part of the magma chamber of a water-bearing basaltic melt that underwent a process called **crystallization differentiation**. The denser olivine and pyroxene crystals settled fastest through the melt to accumulate and form the dark colored rocks while lighter-colored and less-dense plagioclase was left behind as the dark minerals settled downwards. The results of this process can be seen in the low outcrops starting just downslope (east) from the pullout. The first rocks encountered are dark grayish gabbro with olivine, amphibole, and whitish plagioclase. The olivine can be recognized on partially weathered surfaces as small (1–3mm) rounded grains with a reddish coloration due to the oxidation of its iron component in today's surface environment. In fresh, broken pieces, the olivine is lustrous and olive green in color. Amphibole is shiny and black, and crystallized last to form irregular intergranular crystals. Plagioclase is white and commonly forms rect-

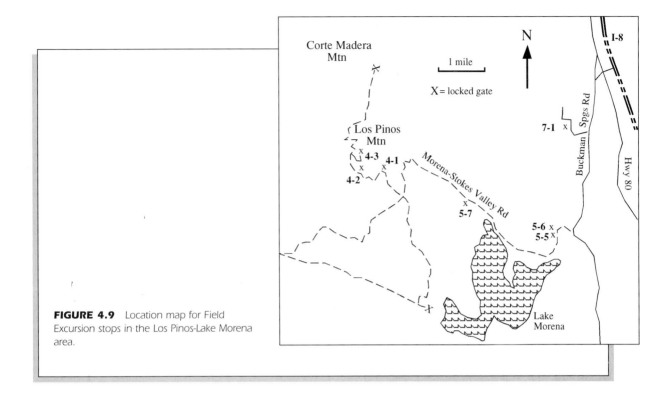

FIGURE 4.9 Location map for Field Excursion stops in the Los Pinos-Lake Morena area.

angular crystals. Walk downslope a few yards further and the rock begins to get darker as olivine and black amphibole become more abundant. The white plagioclase gets scarcer as the rock grades from gabbro into peridotite, a change due to the settling of the dark minerals towards the bottom of the magma chamber.

The white anorthosite further downslope probably represents a fresh batch of melt injected into the magma chamber. If we were to drill underneath those outcrops, the rocks would get darker and darker with depth until peridotite, representing the bottom of that chamber, is encountered. This suggests that the original magma chamber was repeatedly filled with multiple injections of melt, a pattern similar to that observed at the currently active summit of Kilauea in Hawaii.

COMB LAYERS. Continue about 1 mile toward the summit along the gravel road to a sharp left turn with dark outcrops above a small gully on the right-hand side of the road. The outcrops are part of a zone of **comb layers** that surround part of a linear belt of magma chambers (Figure 4.6) that includes the previous stop. The flat surface of the outcrop above the road shows a near vertical set of layers of alternately coarse-grained and sparser fine-grained hornblende gabbro. The thicker, coarse-grained layers have a curious structure called **comb layering** in which elongate hornblende crystals up to several inches in length point in a common direction inwards (uphill) towards the center of the old magma chamber. The fine-grained layers have equidimensional crystals of hornblende and plagioclase (Figure 4.8). Upslope and to the right of the small gully are additional outcrops of comb-layered gabbro that can be reached with some difficulty through the thick brush.

 This layering is thought to have been formed as water-rich basaltic melt rose through a conduit zone represented by the elongate belt of olivine gabbro. The presence of water as a separate fluid in any silicate melt system results in the creation of large crystals (pegmatitic texture) because ionic diffusion rates are much higher in this low viscosity, water-rich fluid and crystals can grow many times larger than would be possible in a silicate melt (see also Chapter 6). The comb layering is thought to occur at a melt-solid interface where exsolved water is rising upwards through the melt towards the surface. The hornblende grains grow faster into the slowly rising stream of water bubbles, pushing the solid-melt interface further into the chamber. The different layers of coarse-grained gabbro then represent growth in relatively quiescent periods when the chambers were full. The fine-grained gabbro layers between the coarser comb layers could have formed when the melt vented at the surface during a chamber infilling stage. This would abruptly lower the total pressure, a phenomenon which leads to rapid crystallization and the observed fine-grained rocks. The multiple episodes of melt injection observed at Stop 4-1 are also indicative of a magma chamber that vented to the surface. Imagine that you were at this point about 120 million years ago when this system was active. You would be standing inside a basaltic volcano watching the melt streaming by on its way towards a surface that was probably several miles above you.

If you have driven to this point, it is well worth your time to finish the drive to the top of the mountain and the fire tower.

 **SUMMIT AREA.** As you reach the summit area, the few smallish pine trees with blackened trunks are the survivors of the great 1970 fire that many long-time residents of San Diego County will remember. One story told to me many years ago came from the gentle lady who stood duty in that fire tower. As the fire progressed, she continued to hold her ground and send out information on its path until it seemed sure to overtake the tower. At this point, she ran to her small car and drove madly down the northwest side of the mountain towards the Corte Madera ranch. This road today is barely passable with a 4-wheel drive vehicle. As she drove out, the front of the fire, whipped by Santa Ana winds, passed her twice. She plunged ahead past the flames each time and escaped to Pine Valley. A very remarkable woman.

The turn to the fire tower at the summit area may be gated unless it is occupied during the fire season but it is a very short walk past it to the summit. The 360° view is awesome (Figure 4.10). The road through Morena-Stokes Valley and Lake Morena is visible below (see also Stop 5-6). Cuyamaca Peak, another large gabbro pluton can be seen to the north, Tecate Peak and Mexico to the south, and, if the air is clear, downtown San Diego to the west.

FIGURE 4.10 *View northward from near the summit of Los Pinos Mountain. The light-colored ridge is the monzogranite at Corte Madera Mountain, a favorite rock climbing spot in San Diego County.*

The rocks are all coarse-grained gabbro similar to those seen at the first stop and have many interesting features. The area just below the summit now has numerous small pines emerging from the chaparral. They were planted in the years following the 1970 fire and are beginning to make themselves part of this ecosystem. Los Pinos will soon again live up to its name.

Tonalite

Tonalite (see Appendix A) underlies about 50 or 60% of the western zone of the batholith and forms large plutonic bodies covering several tens of square kilometers each with some in excess of 100 km². One of the largest tonalite plutons underlies the area around Alpine and runs along I-8 from near the view stop east of Flynn Springs to the Pine Valley bridge. The dark gray rocks exposed in numerous roadcuts contain abundant, dark, ellipsoidal xenoliths or inclusions up to one-half meter in length, a feature typical of this rock unit (Figure 4.4). The pancake-shaped xenoliths, and the dark minerals in the rock itself, are crudely aligned in a near-vertical pattern. This alignment, somewhat similar to that seen in gneissic metamorphic rocks, is the product of late-stage movement within the magma that forced the solid inclusions and the early formed dark minerals to rotate into a rough parallelism (Stop 5-1).

The mineralogy of the tonalites is relatively simple. Plagioclase and quartz comprise at least 70% of most outcrops and dark minerals such as biotite, hornblende, and pyroxene make up the remainder. Biotite is the softest mineral of the group. It can be recognized by gently scraping the shiny black grains with a knife and observing the tiny, golden-brown flakes that form around the point. Hornblende and pyroxene are both harder than a knife blade but can be distinguished by their respective crystal forms. Hornblende typically occurs as black, rectangular grains up to one centimeter in length whereas the pyroxene exists as tiny (1 to 2 mm) dark reddish stubby grains. Magnetite is a minor but important constituent. It cannot be seen in the outcrop but a small magnet dangling from a string will be drawn to the rock signaling its presence.

The ratio of the dark minerals to one another is quite variable but two main associations can be delineated: biotite with hornblende, and biotite with pyroxene. These biotite-hornblende and biotite-pyroxene tonalites occur both as parts of individual plutons and as separate intrusions. Such associations *cannot* be produced by simple crystal differentiation processes, i.e., crystals separating from the melt and accumulating under the influence of gravity; they must represent separate but similar melts generated simultaneously by melting of a single source region. As will be discussed shortly, the identification of that source rock is critical in reconstructing the geologic events that led to the formation of the tonalite plutons, the most abundant component of the Peninsular Ranges Batholith.

The emplacement age of the tonalitic rocks in western San Diego County is well constrained. Zircon is relatively abundant in tonalitic rocks and a number of precise radiometric ages have been determined from all of the large tonalitic plutons. They do not show any evidence of interaction with the much older North American continent as do the rocks of the Late Triassic-Jurassic CLM arc and must therefore have risen through oceanic rather than continental lithosphere (Figure 3.3). The tonalite radiometric ages fall between 109 and 102 Ma and imply that these rocks were generated in a limited span of geologic time. Given possible errors of up to one million years inherent in the determination of these zircon U-Pb emplacement ages, that time span could be as short as 5 million years. If the subducting plate that is presumed to have been responsible for the creation of the tonalite was active for several tens of millions of years, why were these tonalite melts only generated over such a short span of geologic time? Perhaps the answer lies back a bit earlier in the history of the PRB when the gabbroic rocks were created (see page 45 "Multiple Hypotheses and Multiple Source Regions").

Monzogranites

Monzogranites are igneous rocks that contain both alkali and plagioclase feldspars as well as abundant quartz (see Appendix A) and constitute about 25% of the igneous rocks exposed within the western zone of the batholith. Dark minerals such as biotite, hornblende, pyroxene, and magnetite form only around 5 or 10 percent of the rock. Biotite is the most abundant of these and typically forms elongate, crudely aligned, aggregates less than a centimeter in length. The presence of magnetite can be again determined by observing the attraction of a suspended magnet to the outcrop. Dark inclusions, so abundant in the tonalites, are usually scarce in the monzogranites (see, however, Stop 5-6).

As discussed earlier, these rocks form prominent topographic landmarks throughout the county and are the foundation of such recognizable monoliths as Lawson Peak, Stonewall Peak, and Mount Woodson (Figure 4.5). The plutons are typically small, less than 10 km² in area, and are spatially separated from one another giving the appearance of being unrelated units. In addition, the monzogranites show considerable variation in grain size and degree of mineral alignment, ranging from coarse grained and weakly aligned at the Corte Madera Mountain pluton (Figure 4.10), through medium grained and strongly aligned at Chiquito Peak, to porphyritic and moderately gneissic at the Pine Valley pluton (Stop 5-2). These features simply reflect the variations in cooling history each body experienced as it was emplaced into the Earth's crust. Despite these textural differences, the monzogranites have very similar mineralogical and chemical compositions that are consistent from pluton to pluton. This indicates that most of the monzogranite bodies throughout western San Diego County were formed by partial melting of a common source region and are part of a single magma-generating event.

Based entirely on field relationships, the monzogranite plutons have consistently been reported as being among the youngest plutonic rocks in the western zone of the PRB. Dikes, reportedly of monzogranite composition, crosscut several tonalite and gabbro plutons, and inclusions of tonalite have been found in bordering monzogranite intrusions (Stop 5-5). These relative age determinations, however, have recently been challenged by a series of zircon U-Pb ages from several monzogranite plutons that fall between 119 and 111 Ma. These ages indicate that at least some of the monzogranites are as much as 10 to 15 million years *older* than the surrounding tonalites. In addition, the isotopic data do not indicate any interaction of these plutonic bodies with the older rocks of the North American craton. They, and the tonalites, were emplaced into oceanic crust westward from the leading edge of the North American continent.

△ Multiple Hypotheses and Multiple Source Regions

*G*eneration of the basalt-andesite-dacite volcanic suite in island arcs and their gabbro-tonalite-monzogranite plutonic counterparts in the underlying batholiths is a hotly debated issue. Central to the issue is the recognition that during subduction, the region below the volcanic arc in which magmas are generated can contain different source rocks. These include oceanic crust that has been modified by reaction with hydrothermal solutions at a spreading center, subducted oceanic sediments, mantle peridotite in the overlying lithospheric plate, and finally, continental crust (Figure 3.3). The time frame for the emplacement of the Santiago Peak Volcanics (SPV) and the western zone of the Peninsular Ranges Batholith (PRB) (Figure 1.1) shows that the SPV magmatism preceded but may have overlapped the granitic rocks of the PRB. According to conventional wisdom, the three distinct rock units in the western zone of the PRB, gabbro, tonalite, and monzogranite, are equivalent to the basalt-andesite-dacite suite found in volcanic arcs. These relationships argue that the SPV are remnants of the older volcanic cover of the PRB and therefore are part of the same magma-generating subduction event.

The origins of the three different rock units within the western zone are not well understood, partly because of the limited age data available for these rock units. The composition of gabbroic rocks dictates that they could only form by partial melting of peridotite so that the mantle wedge above a subducting and dehydrating oceanic slab is the most likely source for these melts (Figure 3.3). In many volcanic arcs around the world, gabbroic rocks, and their basaltic volcanic cover, are formed early in the subduction event and can create large volcanic structures, presumably with underlying gabbroic plutons. In the PRB, most of the volcanic cover has been removed by erosion so that the underlying gabbroic plutons are exposed to form today's familiar topographic landmarks.

Hypothesis 1

The tonalite and monzogranite could both be derived by partial melting of the upper part of the subducted oceanic crust, i.e., altered basalt (Hypothesis 1). However, the pattern of known radiometric ages between these two granitic components sets up an interesting problem. How can these two magma types be generated by partial melting *of the same source* within a relatively short span of geologic time? Why is it that monzogranite melts at one time and tonalite melts later? One possible answer involves the gabbroic plutons. If they do have a relatively long history and were generated over many tens of millions of years, then the early portions of the PRB were mainly gabbroic in composition. If these gabbroic roots became thick enough to themselves undergo partial melting, the resultant magmas could have the chemical composition of tonalite.

Hypothesis 2

In Hypothesis 2, the tonalitic rocks are created by partial melting of the base of a thickened hornblende gabbro crust as it is heated beyond its melting temperature by the passage of higher temperature olivine gabbro melts. The monzogranites would still be created by partial melting of subducted oceanic crust as in Hypothesis 1 (Figure 4.11). In this scenario, the monzogranite magmas are created near the top of the subduction plane (Figure 3.3) whereas the tonalite melts form at the base of the volcanic arc (Figure 4.11). Do any of these melts meet one another during ascent and mix? Do they change composition by reacting with older rocks? This working model raises many questions, but it is one that geoscientists must consider as they continue to piece together the early geologic history of San Diego County.

Field Excursion 5

Granitic Rocks of the Western PRB

STOP 5-1

ALPINE TONALITE. *Travel east on I-8 approximately 18 miles from College Avenue in San Diego to the Tavern Road exit for Alpine. Proceed east through town on Alpine Boulevard about 1.7 miles to the first left turn (Victoria Drive). Turn left, pass under the freeway and park in the cul-de-sac opposite the large outcrop. This rock with its abundant dark xenoliths (Figure 4.12) is typical of one of the two tonalite end members described in the text. It is a hornblende- and biotite-bearing tonalite with trace amounts of pyroxene that are visible only with a microscope. Close examination of the outcrop shows that the hornblende crystals are commonly rectangular and crudely aligned into a near vertical pattern. The dark xenoliths are much finer grained than the surrounding tonalite and consist of hornblende, biotite, and plagio-*

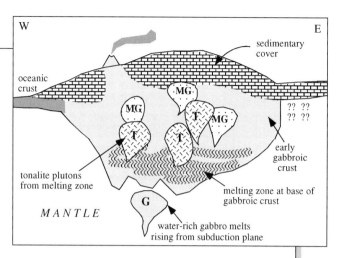

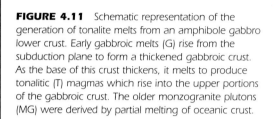

FIGURE 4.11 *Schematic representation of the generation of tonalite melts from an amphibole gabbro lower crust. Early gabbroic melts (G) rise from the subduction plane to form a thickened gabbroic crust. As the base of this crust thickens, it melts to produce tonalitic (T) magmas which rise into the upper portions of the gabbroic crust. The older monzogranite plutons (MG) were derived by partial melting of oceanic crust.*

clase with minor quartz. They are up to a foot or so in length, ellipsoidal, and crudely aligned into the same pattern as the rectangular hornblende grains. This parallelism suggests that the alignment of both the hornblende crystals and the xenoliths is due to upward movement in the magma after emplacement but before final solidification had occurred. The dark xenoliths most likely represent older pieces of the tonalite that had quickly crystallized when the first streams of melt came into contact with the cooler country rocks. Fragments of these finer-grained rocks were then incorporated into new pulses of tonalitic melt that rose into the large chamber and cooled to form the Alpine tonalite. A sample from this outcrop has been dated by zircon U-Pb techniques and gave an emplacement age of 108 Ma.

FIGURE 4.12 Dark inclusions in tonalite along Victoria Drive.

PINE VALLEY MONZOGRANITE. Return to I-8 and proceed east. Take the Sunrise Highway exit and turn left (north) towards Mount Laguna (Figure 2.3). Cross the freeway and turn left (west) onto Old Highway 80 towards Pine Valley. Drive approximately 150 yards to the fresh roadcuts on either side of the road.

Parking is tight at this stop so be careful. Traffic is typically light.

The rock here is a weakly gneissic biotite-bearing monzogranite that contains larger (1 to 3 cm) crystals (phenocrysts) of pale pink orthoclase. Zircons from this outcrop gave an age of 118 Ma that is identical within experimental error to an age of 119 Ma determined on a similar rock about ten miles to the north along Sunrise Highway. This outcrop is a small finger of the larger Pine Valley pluton which forms the buff colored peaks to the northwest. The Pine Valley pluton is unusual in two respects. First, it is the only monzogranite in this part of the batholith with a porphyritic texture. The pale pink orthoclase grains form blocky crystals that are distinctly larger than the remaining plagioclase, quartz, and biotite. This grain-size difference suggests that there were two periods of crystal development, one in which only orthoclase grew from the melt followed by a period in which the remaining minerals crystallized into a mass of smaller crystals. Second, it is one of only a few plutons in the western zone of the batholith that does not contain any magnetite as a test with a small magnet will confirm.

CIBBETT FLAT PLUTON. At this stop, we can examine the biotite- and pyroxene-bearing end member of the tonalite system. Drive 4.5 miles northeast on Sunrise Highway from its intersection with I-8 to the first of several right-hand pullouts. The outcrops opposite the pullout are part of the Cibbett Flat pluton which is exposed over an area of about 80 square km. It consists of a discontinuous, one kilometer-wide rim of hornblende-biotite tonalite similar to that seen at Stop 5-1 and an inner zone of biotite-pyroxene tonalite as seen at this stop.

This outcrop is an excellent example of the weathering style that characterizes the granitic rocks in the PRB. As described in the first part of this chapter, deep chemical weathering during the Miocene was focused along three fracture sets intersecting at right angles. Weathering proceeded inward from the fractures, loosening the individual grains until the solid rock had been converted to a consistency almost like that of loose sand. If the fracture sets were far enough apart, the interior of a large block between fractures remained fresh. When exposed by road construction, these relatively unaltered corestones appear as spheroidal boulders surrounded by "gruss," the deeply weathered remnants of the outer portion of that fracture block (Figure 4.13). In time, erosion will remove the looser material leaving behind corestones piled up like marbles.

FIGURE 4.13 *Corestones in weathered tonalite (gruss) at Stop 5-3 along Sunrise Highway.*

These massive rocks are much darker in color than the Alpine tonalite (Stop 5-1) because the plagioclase feldspar is dark gray in color rather than white. Biotite occurs as irregularly shaped, single crystals up to two centimeters across whose cleavage surfaces can be seen reflecting across the outcrop. Mineral alignment is weak to absent and the dark xenoliths that are so obvious in the biotite- and hornblende-bearing variety are not present. A sample from this pluton just past Crouch Meadows along Sunrise Highway has a reported zircon U-Pb emplacement age of 104 Ma.

STOP 5-4

KITCHEN CREEK ROAD. Continue northward along Sunrise Highway (S-1) for 0.3 miles to Kitchen Creek Road which heads southward for 13 miles until it meets I-8. If the access gate is open, this paved but rough route will take you through the western side of the Cibbett Flat pluton, dropping from the pine-forested Laguna Mountains into the riparian environment near the Cibbet Flat campground and out to I-8 through a series of mountain meadows.

*S*top at a wide gravel shoulder after driving about 0.9 miles along Kitchen Creek Road. To the northwest is the familiar dark outline of Cuyamaca Peak (gabbro) with the light-colored bouldery outcrops of Stonewall Peak (monzogranite) just to its right. Small bouldery outcrops of the biotite-pyroxene tonalite occur all along the road but continue for about 4.1 miles through the "Penny Pines" area to the first long roadcut along the right (west) side of the road. The fresh corestones emerging from the gruss are biotite and pyroxene tonalites similar to those at Stop 5-3. This outcrop is also cut by a sinuous, two-foot thick granite dike. Two subparallel shear planes (faults) may exist here. The first is only a few inches thick and just below and roughly parallel to the dike. Close examination of the granite dike where it comes down to road level shows a rock cleavage and some banding defined by thin, alternating dark- and light-colored layers. This suggests that the dike was emplaced into an active fault that continued to move as the dike solidified.

Continue southward for another 0.7 miles to another long, deeply weath- ered roadcut with a one-meter thick pegmatite dike (see Chapter 5) cutting nearly vertically through tonalite gruss at the south end of the roadcut. This pegmatite shows mineral alignment and zoning characteristic of these com- plex but intriguing rock bodies. Pull over on the dirt shoulder about another one-half mile ahead to view a large light-colored granite (pegmatitic?) dike across the canyon. This approximately one-half mile long, sheet-like feature dips westward towards you at an angle steeper than that of the canyon wall. Where the dike encounters small ravines cut into the canyon wall, the outcrop pattern appears to form a "V" pointing downhill.

Continue ahead another 0.4 miles to the gate and a return to a well-main- tained paved road. The Cibbet Flat Campground is another 0.7 miles ahead. The campground entrance marks the eastward extent of the Cibbet Flat tonalite. Beyond this point, the road cuts through older granodiorite gneisses of the CLM arc (see Chapter 2). Stop 1-4 is about 2.5 miles further ahead. At I-8, turn back to the north (right) and drive about three miles to the Buckman Springs road exit.

CONTACT ZONE. Exit at Buckman Springs Road. Cross Old Highway 80 and continue south on Buckman Springs road approximately 2.3 miles to the Morena-Stokes Valley Road (Figure 4.9). Turn right (west) and drive about 0.6 miles to a small pulloff on the right (west) side of the road in a small grove of oak trees. The dark weathered outcrop adjacent to the pulloff is a hornblende-biotite tonalite similar to that seen at Stop 5-1. A sample of this tonalite from outcrops about 200 yards to the west has a reported zircon U-Pb age of 102 Ma. Examine this outcrop and note the abundance of dark minerals, the relatively coarse grain size, and moderately developed mineral alignment.

Low outcrops in a small gully just to the right (east) of the tonalite out- crop expose the contact zone between the darker, coarse-grained tonalite and a finer-grained biotite-bearing monzogranite body (Figure 4.14). At the

FIGURE 4.14 Contact zone between a coarser-grained, darker tonalite and a younger monzogranite. The held pen is pointing to a foot-sized fragment of tonalite surrounded by finer-grained monzogranite. The marking pen in center is in the monzogranite and oriented parallel to the flow banding in that rock unit. Both features indicate that the tonalite is the older of the two rock units.

contact, inclusions of the coarser grained and darker tonalite can be found surrounded by a finer-grained, lighter-colored monzogranite. The monzogranite in parts of the contact zone exhibits a faint flow banding, and has narrow fingers that protrude into the darker, coarser grained tonalite. These three observations can be used to determine the age of the monzogranite relative to the tonalite and are consistent in suggesting that the monzogranite is the younger of the two, i.e., less than 102 Ma. This makes it the youngest rock found to date in the western zone of the batholith.

A SHORT WALK TO A FASCINATING OUTCROP. *Cross the small gully and hike north towards the light-colored ridge just ahead. In doing so, note that the grain size of the monzogranite increases as you go away from the contact zone, a change again consistent with a younger monzogranite melt partially quenching against the cooler and older tonalite body. Walk through the sparse brush across a series of flat outcrops towards the large, light-colored steep face about 100 yards ahead. There are no trails but it is a relatively easy walk. However, there are rattlesnakes in this area, so observe caution at all times. As you walk across the flat outcrops, observe what appear to be swirls in the monzogranite. This pattern is most likely due to a surge of new material into the magma chamber which distorted the already solid but still plastic carapace of the pluton. Climb carefully up the left-hand side of the large steep face ahead using an existing horse trail and walk over to the center of this very large sheet of rock.*

*F*irst, enjoy the view of Lake Morena off to the south. This reservoir is part of San Diego's water supply. From the dam on its southwestern side, it drains into Cottonwood Creek and down into Barrett Lake. A county park surrounds most of the reservoir and is a popular fishing and camping spot.

The first thing one notices in the outcrop are the numerous dark xenoliths. These are similar to the dark xenoliths seen in many of the tonalite outcrops and probably are pieces of older rock ripped off as the magma moved upwards. Most are elongate and aligned parallel to one another. Some are oddly contorted and bent into folds but still with the same crude alignment. Close examination of these inclusions shows that they are relatively fine grained and porphyritic with small but distinctly larger laths of whitish plagioclase feldspar in a finer-grained dark matrix. This texture indicates that they underwent two periods of cooling. First, plagioclase nucleated and grew in the melt to form the rectangular crystals. Then, the melt was quenched in such a way as to create abundant small crystals of dark minerals such as biotite, hornblende, and minor magnetite. Finally, they were somehow broken away from their parent rock and incorporated into the rising monzogranite melt where they were rafted around like pieces of flotsam.

Now examine the thin white streaks in the outcrop surface. These are small veins of light-colored granite that intruded the monzogranite host. The veinlets contain gray quartz, whitish albite feldspar and pinkish orthoclase. They represent fractures that formed along the margin of the magma chamber when more magma entered the chamber from below, causing it to balloon and split. The fractures were then filled with the remnants of the melt remaining in the top of the old chamber, a process not unlike squeezing the bottom of a toothpaste tube and having your toothbrush fill with the material from the

FIGURE 4.15 Deformed light-colored granitic veinlets and dark inclusions in monzogranite. The parallelism of the axis of folding, stretching direction of the dark inclusions, and alignment of the biotite in the monzogranite host, indicate that all three components acted plastically and must have been at high temperatures during deformation. Note lens cap for scale.

top. These veins were later bent into a series of chevron-like folds (Figure 4.15) as the plastic outer skin of the magma chamber underwent even more movement when the original chamber swelled further due to the influx of more fresh melt. The axes of these folds mimic the alignment of the dark inclusions and argues that both were aligned during the last filling stage of the magma chamber.

STOP 5-7

VIEW STOP ALONG MORENA-STOKES VALLEY ROAD. Return to your vehicle and follow Morena-Stokes Valley Road past the Morena Honor Camp where the surface changes to gravel. Continue along the north side of Morena Reservoir for another 1.2 miles to where the road crosses a small stream feeding the northern end of the lake. About one-half mile after crossing the small stream, pull over along the meadow that affords a clear view of the large ridge to the right (east) of the road.

The grayish blocky rock is tonalite similar to that seen at Stop 5-1. It is part of a large tonalite pluton that includes all of the surrounding low rolling hills. The buff-colored bouldery outcrops are monzogranites that may be part of the small pluton seen at the previous stop. The dark, reddish outcrops are quartzites and biotite-quartz gneisses of the Julian Schist that form a **roof pendant** or package of older metamorphic rocks engulfed by and entirely enclosed in younger igneous rocks. This relationship can also be viewed from the top of Los Pinos Mountain (see Stop 4-3). A narrow band of whitish to buff-colored rock cuts across the darker quartzites. This feature is a dike, i.e., a sheet-like body of igneous rock that was forced into a fracture within the older metamorphic rocks.

△ The Eastern Arc: Consistency and Quiescence

The rocks of the eastern zone of the PRB reflect conditions of magma generation, emplacement, and cooling that were very different from those in the western zone. They were created following a rapid eastward shift in the position of the Cretaceous volcanic arc. Igneous activity in the western zone came to an abrupt halt just prior to 100 Ma and shifted eastward several tens of kilometers to create a series of very large plutons at about 95 Ma. In effect, the "faucets" that were feeding the plutons (and overlying volcanoes) in the west were turned off and, after a short hiatus, new ones well to the east were turned on. This eastward jump of the Cretaceous arc is even more amazing when one recognizes that it is not restricted to San Diego County. This event is evident in all of the eastern Peninsular Ranges from Riverside County southward through most of Baja California and must therefore reflect dramatic changes in plate motions beginning at about 100 Ma.

If one gathers a handful of weathered debris from around almost any outcrop of plutonic rock in the western zone of the PRB and runs a small magnet through the loose material, small bits of dark minerals will jump up and adhere to the magnet. These dark, opaque mineral grains are magnetite, an oxide of iron and one of a very few minerals that is strongly magnetic. It makes up only about one to two percent of any plutonic rock in the western zone. In the eastern zone, similar weathered outcrop material will *not* contain any magnetite. The bits of dark, opaque minerals found in these rocks are ilmenite, a nonmagnetic oxide of iron and titanium. This mineralogical difference is directly related to the availability (activity) of oxygen in the source rocks that were melted to create each zone and implies different sources and/or conditions for each plutonic zone. This seemingly minor change from magnetite in the west to ilmenite in the east, when plotted on a geologic map of Southern and Baja California, runs the length of the PRB and divides it into two discrete zones (Figure 4.1). This change also coincides with the boundary between the older western and younger eastern zones, and thus represents a fundamental change in the conditions that gave rise to the plutonic rocks of the eastern PRB.

△ The La Posta Pluton

The eastern zone of the PRB from San Diego County southward through Baja California is dominated by a series of large plutons that have nearly identical emplacement ages, rock types, and physical characteristics. The largest of these, the La Posta pluton, underlies most of eastern San Diego County and covers nearly 1400 km². It stands in stark contrast to the smaller plutons (<< 100 km²) of the western zone. What makes this body of rock so phenomenal is that it represents a *single pulse* of magma that slowly cooled inward to form a series of concentric rock units (Figure 4.16), a feature typical of all of the La Posta-type plutons in this zone but atypical of plutons in the western zone. The sequence of rock units going from the rim of this pluton inward to its core record conditions of undisturbed cooling with inward migration of the crystallization front and make this pluton a textbook example of *crystallization differentiation.*

Although silicate magmas contain a complex mixture of elements such as aluminum, iron, magnesium, titanium, potassium, calcium and sodium, the main chemical constituents are silica and oxygen. These two elements combine to create the basic building block of silicate minerals, the silica tetrahe-

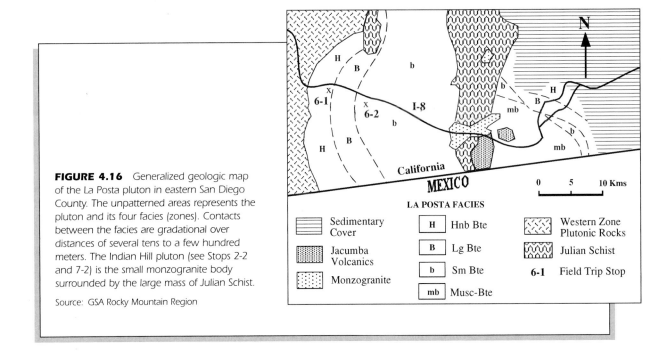

FIGURE 4.16 *Generalized geologic map of the La Posta pluton in eastern San Diego County. The unpatterned areas represents the pluton and its four facies (zones). Contacts between the facies are gradational over distances of several tens to a few hundred meters. The Indian Hill pluton (see Stops 2-2 and 7-2) is the small monzogranite body surrounded by the large mass of Julian Schist.*

Source: GSA Rocky Mountain Region

dron (Figure 4.17). The tetrahedra combine with the other elements to make different minerals. The specific minerals that form are a function of the composition of the magma and the temperature and pressure (depth) at which crystallization takes place. Minerals crystallize in specific order and over a wide range of temperatures as the magma cools from a completely molten state to the last few remaining ribbons of melt woven into a solid mass of newly formed silicate minerals.

The first minerals to form from the parent melt are different from those that crystallize at the end of the sequence. Thus, as the early minerals form, the composition of the remaining (daughter) melt changes so that new and different minerals will begin to nucleate and grow. This process by which the daughter melts continuously change composition through crystallization is known as **crystallization differentiation**. Melts of granitic composition are richer in silica and thus much more viscous than melts of gabbroic composition. Rather than sink down into this magma, as was observed in the gabbroic magma chamber at Los Pinos Mountain (Stop 4-1), early formed crystals remain more or less in place and cause the front of crystallization to slowly migrate inward. The adjacent slightly cooler daughter melt sinks slowly into the interior of the magma chamber. Hotter melt from the interior rises to take its place. The daughter melts mix in the deeper and hotter parts of the chamber and begin to rise anew. This motion, taken in whole around the periphery of the chamber, creates convection cells that force daughter melts to change composition as the crystallization front slowly migrates inward towards what is now the core of the pluton.

The large La Posta pluton and its sister (La Posta-type) plutons up and down the length of Southern and Baja California are all concentrically zoned with

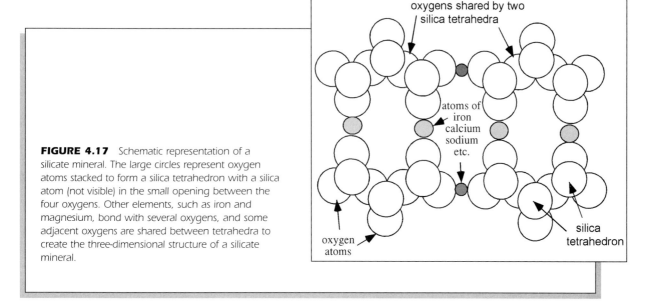

oxygens shared by two
silica tetrahedra

atoms of
iron
calcium
sodium
etc.

oxygen
atoms

silica
tetrahedron

FIGURE 4.17 *Schematic representation of a silicate mineral. The large circles represent oxygen atoms stacked to form a silica tetrahedron with a silica atom (not visible) in the small opening between the four oxygens. Other elements, such as iron and magnesium, bond with several oxygens, and some adjacent oxygens are shared between tetrahedra to create the three-dimensional structure of a silicate mineral.*

progressive inward changes in mineralogy that could only occur within slowly cooled magma chambers that were not disturbed by either tectonic forces or large influxes of fresh melt. The outer zone of the pluton (Figure 4.16) is a tonalite that contains sharply rectangular hornblende crystals up to one centimeter in length along with smaller hexagonal grains of biotite, and honey-colored, wedge-shaped sphene, a calcium-titanium silicate, all set into a matrix of white plagioclase and grayish quartz (Hornblende-Biotite Facies). The next zone inward (Large Biotite Facies) is a granodiorite characterized by a general absence of large rectangular hornblende crystals and an increase in size of biotite hexagons to about one centimeter. Alkali feldspar is now present in small amounts but its whitish color makes it difficult to distinguish from plagioclase. The third zone (Small Biotite Facies) is a monzogranite that contains abundant plagioclase, quartz, and alkali feldspar but no hornblende. Biotite still exists as hexagonal crystals but its grain size has diminished to a few millimeters. The rocks in the core of the pluton (Muscovite-Biotite Facies) have the same mineralogy and grain size but can be distinguished from the third zone by the presence of tiny flakes of muscovite, a colorless mica. While these changes may seem minor, they reflect the inwardly changing composition of the melt that concentrated the late-forming, potassium-rich minerals (alkali feldspar and muscovite) in the core zone. The relatively stable conditions under which this magma chamber crystallized are also reflected in the gradational nature of the contacts between adjacent rock units (see, for example, Stop 6-1). The lateral distance from one clearly recognizable unit to the next typically ranges from several tens of meters to several hundred meters. In that distance, the rocks exhibit very subtle changes and appear to blend gradually from one unit into the next without a well defined point of change. Such contacts could only occur if the melt remained undisturbed as it underwent convection and inward crystallization.

Uplift and Erosion

The awesome size of the La Posta pluton ($\sim$1,400 km^2) seems even more remarkable when one considers that similar plutons were being emplaced at essentially the same time in a sweeping arc that reached from Riverside County southward through the State of Baja California. Unlike the arc that gave rise to the plutons of the western PRB, this system was relatively short-lived. Its magmatic activity began between 98 and 95 Ma and was likely over, at least in the greater San Diego County area, before 90 Ma. In that geologically brief time, the volume of rock added to the North American continent is roughly equal to that formed in the 20 or 30 million year history of the western PRB.

Imagine yourself standing *nervously* on the east-facing flanks of a dying western-zone volcano 100 million years ago. To the north and south, smoldering volcanic landforms attest to the colossal forces still at work within the western zone. To the east, off the flanks of these composite volcanoes, are shallow seas whose sedimentary record has since been stripped away by the uplift and devastation that would follow. The seas stretch perhaps only a few tens of kilometers to the distant landmass of the ancient North American continent. If we could compress time, what a geologist does best, the seas would drain away and rock masses would be forced upwards as the La Posta-type magmas were emplaced nearly 15 kilometers below. The land surface rose to accommodate the newly added volume of magma and blotted out views of the distant continent. Great faults opened passageways into the underlying magma chambers and huge volumes of ash and pumice blanketed the area. The area continued to rise and a mountain chain that would rival the modern Sierra Nevada altered wind and weather patterns. Erosion could not keep pace with the rapid uplift and oversteepened slopes began to fail. Great block-like masses of rock and debris, perhaps the size of the Laguna Mountains, slid away to reveal the tops of now congealed magma chambers. Rain and surface waters penetrated through the cool fractured rocks into the hotter zones above the still solidifying magma chambers. Heated by contact with these rocks, the meteoric waters reached temperatures near 400°C and rose back towards the surface through the same fracture systems. As they did, the rocks were altered in much the same way that oceanic basalts are altered at spreading centers. These solutions, however, did not emerge into a cold ocean floor to immediately cool and precipitate their dissolved load. Some found their way to the surface where the hot waters boiled into steam and created geysers. Others cooled within a few hundred meters of the surface to form hot springs with underlying veins of lead, zinc, copper, gold and silver, a surficial environment not unlike that of modern-day Yellowstone National Park or the North Island of New Zealand.

How does such a change come about? What caused the magmatism to shift abruptly eastward and why were such large volumes of melt formed? These perplexing questions find partial answers in the reconstructed motions of lithospheric plates. As one plate is forced to subduct beneath another, the melting within or above the underthrust plate is, in part, controlled by the rate and direction of subduction. Prior to about 115 Ma, the North American plate was moving northwestward at about two inches per year while the adjacent Farallon plate, precursor to the current Pacific plate, was sliding north at about the same rate. This oblique collision created the northwest-striking arc of the western zone of the PRB. Between 115 and 85 Ma, a portion of the Farallon plate split away and began to move more rapidly into the North American continent. This shift roughly coincides with the change from western to eastern zone magmatism. With a greater rate of subduction, the subducting plate shallowed and magmatism shifted eastward (Figure 4.18) leaving the western zone volcanoes high, dry, and on their way towards extinction.

There are many things to consider about the La Posta pluton and its sister bodies scattered throughout Southern and Baja California. For example, what effect did the rise of the magma have on the overlying rocks? Were they just shouldered aside or were they pushed upwards like pistons to form the peaks of an ancient mountain range? Part of the answer lies in the degree of metamorphism that the older sedimentary rocks, now the Julian Schist, have experienced. The metamorphic minerals in the metasedimentary rocks in the western zone of the batholith record temperatures as high as 600°C and pressures (depth of burial) of 9 km (5.4 miles). To have these rocks exposed at the surface today requires that they (and the rocks above them) be uplifted that same vertical distance. In the eastern zone, however, this same rock system reached 640°C and 12 km (7.2 miles). To have these rocks exposed at the surface today requires the eastern zone to have been raised 3 km (1.8 miles) *higher* than the western zone. At their maximum elevation, this mountain range would rival anything we see today in North America.

Towering tens of thousands of feet above sea level, hemispheric wind and weather patterns would have been disrupted. A "rain shadow" similar to what we see in the Pacific Northwest today could easily have been created so that for millions of years, the lower elevations on the western slopes of the ancestral Peninsular Ranges could have experienced rainfall amounts approaching an inch per day. Warmer temperatures, decayed vegetation, and increased carbon dioxide in the atmosphere raised the pH of groundwater that filtered into and chemically weathered the solid rock. This intense chemical disintegration combined with the physical relief certainly contributed to the destruction of the ancestral Peninsular Ranges.

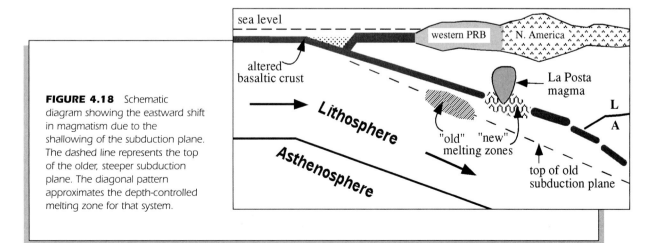

FIGURE 4.18 *Schematic diagram showing the eastward shift in magmatism due to the shallowing of the subduction plane. The dashed line represents the top of the older, steeper subduction plane. The diagonal pattern approximates the depth-controlled melting zone for that system.*

When did this uplift occur and how rapidly were the mountain peaks stripped away by erosion? Evidence has trickled in over the past few years suggesting a most intriguing scenario. The main uplift stage, as would be expected given the size of the plutons, appears to have occurred shortly after the emplacement of the La Posta magma (94 Ma). The overlying rocks were lifted in a crude, piston-like fashion by the rising magma to create the ancestral Peninsular Ranges. By approximately 75 Ma, however, the La Posta rocks, which are currently exposed at the surface, had cooled to temperatures of about 100°C and could only have been at depths of about three or four kilometers. Thus, between 94 and 75 Ma, an interval of about 20 million years, the core of the ancestral Peninsular Ranges was uplifted about 10 km and then stripped down by erosional forces to something only a few kilometers or so higher than what we see today. The central question, however, remains unanswered. How do towering mountain ranges rise and get decapitated in a span of 20 million years only to be followed by considerably less erosion in the subsequent 75 million years? To add to this Holmesian mystery, the sedimentary products of much of this event appear to be missing. If erosion via running water or glaciers destroyed the ancestral mountain range, then the eroded material derived from the ancient highlands should reside nearby as sedimentary rocks of somewhat younger age in a mountain-fringing sedimentary basin. The oldest series of sedimentary rocks in western San Diego County are conglomerates and sandstones that were deposited between 80 and 71 Ma. Upon close examination, the rock fragments in these conglomerates bear little resemblance to plutonic rocks in eastern San Diego County and may represent only the latter stages of erosion of the western side of the ancestral Peninsular Ranges. Where are the rocks that were stripped from the top of the La Posta pluton? Are the eroded materials buried somewhere offshore under even younger sediments? Could they have slid as massive, mountain-range-sized blocks away from the high points of the mountain range? Could ancient faults similar to the San Andreas or the Elsinore have moved these blocks and their associated sedimentary material elsewhere, perhaps (as has been suggested by Canadian geologists) even as far away as British Columbia? There is much yet to unravel.

Field Excursion 6

The La Posta Pluton

A Gradational Contact. *A series of outcrops along Old Highway 80 just east of the small community of La Posta (Figure 4.19) exposes the gradational contact between the outer Hornblende-Biotite Facies and the Large Biotite Facies (Figure 4.16). Take I-8 east approximately 45 miles from San Diego to the La Posta-Kitchen Creek exit. The western edge of the La Posta pluton cuts across the freeway at the off-ramp. Turn east (left) onto Old Highway 80 and travel through town (don't blink) about one-half mile to the first set of outcrops. Park in the wide area on the left (north) side of the road. Traffic is usually light but observe caution at all times.*

The freshest exposures in the outcrop occur near the west side of the roadcut. The rock is tonalite with rectangular black hornblende, hexagonal black biotite, and small amber to honey-colored sphene. Some of the rectangular dark shapes in the outcrop are not hornblende but biotite grains

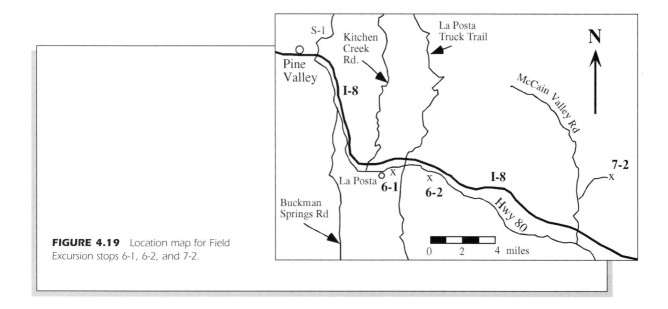

FIGURE 4.19 Location map for Field Excursion stops 6-1, 6-2, and 7-2.

turned on their side exposing a rectangular rather than hexagonal cross section. These can be distinguished from hornblende by gently scratching the grain with a knife blade or other sharp, pointed tool. The biotite is much softer and will yield small, brownish flakes. This outcrop has been mapped as part of the marginal **Hornblende-Biotite Facies** (Figure 4.16).

Examine the outcrop from about three or four feet away. Notice that there are a few dark inclusions up to a foot or so in length and that they have a crude, vertical alignment. Identify several of the rectangular hornblende crystals and compare their orientation with that of the inclusions. In general, the crystals and the inclusions are aligned in the same direction. This weak but visible alignment is most likely due to upward flow of the magma just after the hornblende crystals formed but well before the melt had completely solidified. Drive or cautiously walk about a hundred meters east to another wide parking area on the left (north) side of the road. Examine the outcrop in the same manner and compare notes. Do the dark minerals occur in the same proportions and in the same sizes? Is the mineral alignment still apparent? Are the dark inclusions as abundant? Again, drive another hundred meters or so to the end of the series of outcrops and repeat your observations. When finished, drive about one-fourth of a mile to the intersection of Old Highway 80 and the La Posta Truck Trail. The high roadcut ahead on Old Highway 80 is very weathered but the rocks are all part of the second zone in the pluton, the **Large Biotite Facies**. Examine any of these rocks or turn left (north) and park by a small outcrop about a hundred feet or so from the intersection. This rock has no foliation (mineral alignment), little or no hornblende, and abundant, large (1 cm), hexagonal biotite grains. Now, with your observations from the previous stops, where would you draw the contact between the two rock units? The first person to map this area, Joann Kimsey, a graduate student at San Diego State University in 1985, arbitrarily placed the contact in the river valley behind you. Others have placed it nearer to the third outcrop because that stop had abundant, centimeter-sized biotite grains. This traverse illustrates nicely the difficulties faced in mapping gradational contacts.

THE SMALL BIOTITE FACIES. *Continue driving east on Old Highway 80 for about 3.2 miles to a large set of very fresh outcrops on the left (north) side of the road. Park well off the paved surface on the right. At this stop, we are close to the geographic center of the pluton and as you look to the south, everything that you can see is within the La Posta pluton.*

ook back at the outcrop. If the sun is shining, you may be able to see bright flashes of reflected light coming off the outcrop. The light is bouncing off cleavage surfaces in the mineral orthoclase, an alkali feldspar. If you closely examine one of these feldspar surfaces, it will have the same whitish color as the plagioclase so that except for its size, the two feldspars cannot be easily distinguished. These larger alkali feldspar crystals represent the last mineral to crystallize from the daughter melt that solidified to create this rock. Close examination will reveal that they surround grains of the early formed minerals such as grayish quartz, dark biotite, and rectangular whitish plagioclase. Hornblende is absent and the biotite grains in this outcrop are clearly smaller than those in the Large Biotite Facies. In addition, and as noted above, these rocks contain more alkali feldspar so that this rock falls into the monzogranite field (see Appendix A, Figure A-1).

The interior **Muscovite-Biotite Facies** of the La Posta pluton is nearly identical to the rock seen at this outcrop except for a modest increase in alkali feldspar and the presence of small flakes of muscovite. It can be seen at any of the outcrops off the Mountain Springs exit ramp.

As one continues eastward along Interstate 8, dropping down the grade towards the low desert, the units described above appear in reverse order (Figure 4.16) until, just before the exit to Calexico, the last rocks along the freeway are back in the **Hornblende-Biotite Facies**.

Stopping along the freeway is dangerous and not recommended.

CHAPTER 5

GEMS AND GOLD

⌂ Gems

*A*s any visitor to Julian, California, knows, the discovery of gold has played an important role in the economic and cultural development of San Diego County. However, another resource, perhaps less well known to the casual tourist, has made our area world famous. Tourmaline, a complex boron-bearing silicate mineral, is found in a variety of geological environments but none more rewarding than in rock units known simply as pegmatites. **Pegmatite** is a term for *any* very coarse-grained igneous rock but the term is most commonly applied to igneous rocks of granitic composition whose most abundant minerals are feldspar, quartz, and mica. By strict and arbitrary definition, pegmatites have an average grain size in excess of three centimeters. Some of these rocks, however, have single grains that exceed one meter in size. They are also noted for their "pockets," open spaces in the rock that may be partially filled with clay, mineral specimens of unusual composition (some of which have superb crystal form), and gems, i.e., colored, translucent minerals that can be faceted or shaped to create aesthetic objects of commercial value. From 1898 to 1914, the Himalaya mine near Mesa Grande was the world's leading supplier of tourmaline. Most of it went to Imperial China, where it was used for carving, until the Manchu Dynasty was overthrown in 1912. Today, visitors to museums in Beijing can view countless artifacts carved from San Diego County tourmaline. When these colored minerals also have superb crystal form, they become highly prized by mineral collectors and their commercial value increases dramatically.

The first recognition of gem minerals in San Diego County came in the 1880s when noted mineralogist George Kunz is said to have observed Indian children playing with colored marbles near the Mesa Grande store. Legend has it that he purchased these "toys" for pennies and had them identified by Tiffany in New York as gemmy cores of stream-worn tourmaline crystals. Early production of gem minerals, especially tourmaline, from San Diego County pegmatites was labor intensive and only a few hardy and persistent souls managed to eke out a living as gem miners. Some of the recovered material was cut locally into gemstones but most of it was shipped off to Imperial China. After the loss of the tourmaline market in China, gem production in the county was limited until the Himalaya Mine was reopened by Ralph Potter in 1952. It has been run since 1977 by Pala International and is one of the world's major suppliers of colored tourmaline and other gem materials.

Pegmatites

A pegmatite forms when a *water-rich*, silicate magma is injected into a pre-existing fracture. The presence of abundant water dissolved in the original or parent magma is a critical factor in forming the large crystals so characteristic of pegmatites. Many geologic textbooks still recite the ancient litany that pegmatites form because the magma underwent very slow cooling so as to allow crystals more time to grow. It seems unlikely, if not impossible, that these bodies in San Diego County and elsewhere cooled any slower than their surrounding host rocks. Calculations on the cooling rates of other pegmatite systems in North America indicate that even bodies several tens of meters in thickness cooled fast enough that they solidified in a matter of tens or hundreds of years, an instant in terms of geologic time. Thus, the large crystals owe their size to something else and that something else is water.

If a water-rich silicate melt is forced upwards into fractures to begin solidifying, feldspar and quartz are among the first crystals to form. Both are anhydrous (water-free) minerals so that the remaining daughter melt is even further enriched in water because there is less melt but the same number of water molecules dissolved in that melt. This process, crystallization differentiation, continues as the melt solidifies inward until it becomes saturated in that component, i.e., until the remaining melt cannot hold any more water in solution. Now something strange begins to happen. The melt begins to *boil,* not because it has been heated but because it continues to *cool.* Vapor "bubbles" begin to form, grow, and slowly coalesce into larger "bubbles." As the magma continues to cool inward from its margins, the extra vapor begins to collect in the interior of the dike and the conditions necessary to form the characteristic large crystals in granite pegmatites have been met. Experimental studies have shown that crystals in contact with a vapor grow hundreds of thousands of times faster than in the more viscous silicate melt. Thus, the largest and most desirable crystals in pegmatites are found in the interior of the body where the vapor phase has collected during inward crystallization.

This process also leads to a crude inward zoning in pegmatites (Figure 5.1) that can be used as a guide to finding where the pockets are most likely to have formed. As the melt crystallizes inward from the cooler country rock into which it was injected, one of the earliest sections to form is a peculiar fine-grained combination of feldspar, quartz, and orange-red garnet. The garnets are small, perhaps a millimeter or two, but form thin, discontinuous, cuspate layers that are referred to as "line rock" (Figure 5.2). Although this material is common in many pegmatite bodies, its origin is unclear. The bulk of the material enveloping the "line rock" and leading into the interior of the dike is an intergrowth of gray quartz and whitish feldspar that is referred to as graphic granite because the intergrowth looks crudely like ancient hieroglyphs. Eventually, the remaining melt becomes so enriched in water that large crystals begin to form where the water bubbles have coalesced. This happens in the middle portion of the dike system where the rising bubbles bump into the downward migrating crystallization front. In this zone, the growing crystals

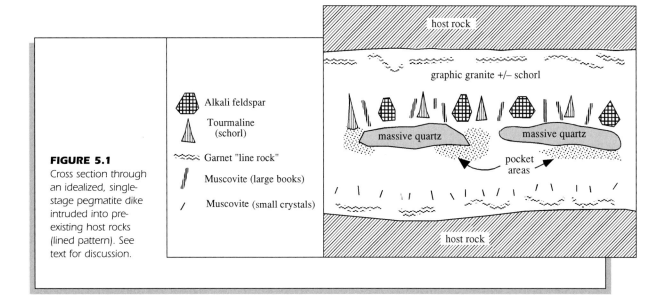

FIGURE 5.1
Cross section through an idealized, single-stage pegmatite dike intruded into pre-existing host rocks (lined pattern). See text for discussion.

Alkali feldspar

Tourmaline (schorl)

Garnet "line rock"

Muscovite (large books)

Muscovite (small crystals)

take on unique shapes in that they broaden downward as a result of the increased rate of crystallization when they come into contact with the coalesced vapor bubbles (Figure 5.1). This particular feature has been used as an indicator that the mining operation may be approaching a pocket area because these crystals taper or point away from the potential pocket areas.

The inner zones of pegmatites may contain a variety of exotic and gemmy minerals. Pink rubellite and green elbaite (both varieties of tourmaline), kunzite (the purple variety of spodumene named after mineralogist George Kunz who first recognized San Diego County as a source of gem material), pink morganite and blue-green aquamarine (varieties of the mineral beryl), all contain rare elements that do not fit into the structure of the quartz, feldspar, and micas that crystallize early from the melt. For tourmaline, the critical element is boron; for spodumene, it is lithium; and beryl requires large quantities of beryllium. As crystallization proceeds, these light elements may become concentrated along with the water in the interior of the pegmatite so that if the late-stage fluids in the dike become sufficiently enriched in any or all of these rare elements, exotic and sometimes gemmy minerals may form.

The process does not end there. The final solidification of the pegmatite is very complex and involves many different physical and chemical processes. Once the silicate melt has finally solidified, a water-rich vapor may still exist. If it leaks out of the pocket area, the pocket may remain as an open space partly filled with well-formed crystal specimens. If it remains, it may react with the enclosing feldspar and alter it to clay. The pocket area may collapse inward to fill with this clay and the crystals and crystal fragments that were lining the pocket interior. If the vapor expands and reaches a critical pressure, it can fracture the crystals and leave the bottom of the pocket littered with their fragments. Pegmatite mining is a very chancy venture. Many pockets opened after careful and labor-intensive efforts have nothing but clay and

FIGURE 5.2 Outcrop view of "line rock" near the base of a pegmatite dike. Note the coarser dark tourmaline (schorl) near the hammer head.

quartz. Others, however, can be rewarding enough to continue mining. Stories of the fabulous gem pockets opened at the Himalaya Mine in the Pala District have become legends and I was fortunate enough to have watched a small pocket opened there in 1988. The mine manager reached into a small space about the size of a soccer ball and extracted a fist-sized mass of red clay. He reached down into a nearby tar bucket full of water and washed the clay around in his hands. After a few seconds, he held up a beautiful, doubly terminated, tricolored tourmaline crystal nearly three inches in length. It had been encased in clay during the final stages of pocket development and had remained there intact for nearly 100 million years.

Where do these water-rich granitic melts come from? Why do they contain anomalous quantities of water? How are they related to the tectonic events that gave rise to the PRB? Pegmatites typically form tabular or sheet-like bodies called **dikes** that were injected as magma into fractures in older rocks. Those dikes that have been worked for gem or specimen material cluster into a series of mining districts (Figure 5.3) that, when superimposed on a geologic map of the Peninsular Ranges, appear to straddle the boundary between the western and eastern zones. Almost all of the pegmatite dikes in Southern and Baja California cut across either western zone plutons or the metamorphic rocks of the Julian Schist and so must be younger than any of those host rocks. Few, if any, have field relationships that indicate the dikes are younger than the La Posta pluton. Radiometric ages between 94 and 99 Ma recently reported on these rocks by geologists from the United States Geological Survey in Denver confirm that they are slightly older than the La Posta pluton (94 Ma) but younger than all of the western zone rocks (>104 Ma). It appears, therefore, that this pegmatite system was created in a very short span of geologic time and is somehow related to the events that gave rise to the eastern zone of the PRB.

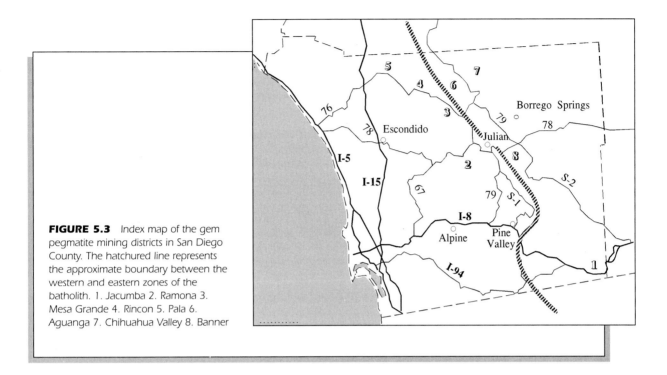

FIGURE 5.3 Index map of the gem pegmatite mining districts in San Diego County. The hatchured line represents the approximate boundary between the western and eastern zones of the batholith. 1. Jacumba 2. Ramona 3. Mesa Grande 4. Rincon 5. Pala 6. Aguanga 7. Chihuahua Valley 8. Banner

As discussed earlier, the metamorphic rocks of the Julian Schist within the eastern zone of the PRB have been subjected to higher temperatures and pressures than those in the western zone. They also have a somewhat different character in that many of these rocks are banded and consist of irregular layers of light-colored rock sandwiched by darker, more biotite-rich rock. The layers swirl and bend much like an incomplete mixing of different colors of cake batter. The light-colored layers contain roughly equal amounts of quartz and feldspar with only traces of other minerals and represent thin layers of granite. The granite, in turn, represents thin streaks of molten material and indicates that this rock had been subjected to temperatures and pressures high enough to begin melting. This mixture of both igneous and unmelted metamorphic rocks is called a **migmatite** (Figure 5.4). The lighter-colored granitic material represents the molten portion of the rock and the darker, biotite-rich material is the part of the rock that had not yet reached its melting temperature, i.e., the unmelted residue. The thicknesses of the light-colored granitic layers vary from one or two centimeters upwards to something approaching one meter. The thicker layers are very coarse grained and, in places, have pegmatitic grain sizes. Thus, the layers and, by inference, the pegmatite dikes were formed by the partial melting of portions of the Julian Schist. Pegmatites are abundant within outcrops of the Julian Schist in the eastern zone of the PRB but are sparse to absent in the Julian Schist exposures within the western zone.

Many of the famous gem-producing pegmatites such as the Himalaya Mine, however, are found in western zone plutonic rocks but near the boundary between the two zones (Figure 5.3). The ages of several pegmatite dikes (95–98

FIGURE 5.4 *Migmatite consisting of light-colored veinlets of quartz and feldspar sandwiched between darker, biotite-rich layers. The quartz and feldspar represent the solidified granitic melt derived by partial melting of the original rock (muscovite schist?, see text) whereas the biotite is the refractory (unmelted) residue. During melting, the muscovite (plus some feldspar and quartz) dissolved together to form the molten portion. The biotite, with its higher melting temperature, remained solid and was concentrated by migration of the melt into the bands.*

Ma) and the lack of dikes within the La Posta pluton suggest that the melting process was initiated when the La Posta pluton began to form and rise into the overlying Julian Schist. Heat energy from this magma body raised the already high temperatures in the Julian Schist beyond the start of melting. One of the first minerals to melt in these rocks is muscovite, a water-rich mica, which was in relative abundance in the metamorphosed shale of the Julian Schist. The water released by the destruction of muscovite dissolves into the newly created silicate melt and lowers its viscosity. As fingers of melt rise through fractures in the metamorphic rocks, some freeze into pegmatite dikes within that host while others seep upwards and laterally into older rocks such as the gabbro and tonalite plutons of the western PRB near the rising La Posta body (Figure 5.5). Once in place, if the water-rich melt crystallizes undisturbed so that the water and other volatiles cannot escape from the magma chamber, then the stage is set for the development of large crystals and, potentially, the formation of the gem and crystal specimens so treasured by collectors.

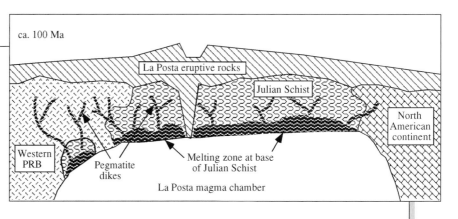

FIGURE 5.5 *Schematic diagram illustrating the transfer of heat energy from the La Posta magma chamber (unpatterned) into the overlying Julian Schist (squiggle pattern). Water-rich pegmatite dikes created by the partial melting of the base of the Julian Schist roof pendants leak upwards and laterally into older host rocks.*

STOP 7-1

Field Excursion 7

Pegmatites and Migmatites

PEGMATITE DIKE NEAR BUCKMAN MOUNTAIN. Take I-8 about 45 miles east from San Diego to the Buckman Springs Road exit. Proceed south along Buckman Springs Road for about 1.2 miles to a gated right (west) turn onto Bear Valley Road. The land behind the gate is public access. The gate is usually closed but unlocked except in times of extreme fire danger. Follow this gravel road, rough in places, for about one mile (Figure 4.9) to a wide parking area atop a low ridge on the left (west) side of the road. Walk southward along the ridge for about 100 meters or so to a series of low white outcrops strung out along the ridge crest. The walk is level but the brush has grown to about two feet, making the walk a bit more difficult. As always, proceed cautiously as this is rattlesnake country.

A s you proceed south along the ridge crest, examine the low, dark outcrops and loose weathered rock along the way. These rocks are similar to the rocks seen on Los Pinos Mountain which is about three miles east of this point. Black hornblende, whitish plagioclase and reddish weathered olivine are the major constituents of this gabbroic body which forms the bulk of Buckman Mountain, the immediate high ridge behind you and to the northwest. This setting, a pegmatitic dike injected into a fracture within a gabbroic host, is identical to that of the renowned gem-bearing dikes near Pala and the world-famous Himalaya Mine near Mesa Grande, California. This common but not exclusive rock association has led some geologists to speculate that the formation of gem-bearing pockets is somehow related to the nature of the enclosing rock, i.e., that only pegmatite dikes cutting through gabbroic host rocks are capable of being gem producers. Stop 7-2, however, as well as gem-producing dikes in the Ramona, California, area cast some doubt on that conclusion.

The dike itself is a relatively small body but this alone does not preclude fabulous riches. The dike at the Himalaya Mine near Mesa Grande has been producing specimen material and gems for over a century and is no more than about one-meter thick. The whitish outcrops contain abundant quartz and feldspar as well as red garnet, black tourmaline and scattered mica. Structures similar to some illustrated in Figure 5.1 can be seen, including modest-sized quartz cores and tourmaline oriented in radial patterns. In places, the black tourmaline occurs as small "broomstraw" clusters. At this point, if you look around at the surrounding low hills to the north and west, you will observe a family of elongate, whitish dikes. Some of these are also pegmatitic in nature but have little in the way of encouraging minerals at the surface.

STOP 7-2

SACATONE SPRINGS. Return to I-8 and travel east about 15 miles to the Boulevard exit (Figure 4.19). Turn right (south) and proceed about one-half mile to the intersection with Old Highway 80. Turn left (east) and drive through town about two miles to the McCain Valley Road. Turn left (north) on the paved surface. In about two miles, the road surface changes to gravel but is well maintained. About one-half mile past the start of the gravel section, turn right onto another gravel road marked with the Bureau of Land Management (BLM) recreational-use sign. This road leads to the

Sacatone Springs overlook and is usually well maintained. It passes through the Small Biotite Facies of the La Posta pluton but after about one mile, a low ridge appears to the left (north). This ridge and the higher topography to the east are underlain by **migmatites** *(see text discussion and Figure 5.4) which are in the western edge of a very complex roof pendant that lies in the middle of the La Posta pluton. A short walk along old trails from one of several pullouts leads to the foot of the ridge where these rocks can be observed firsthand. Park in one of the several small dirt turnouts about one mile from the turn and look for an old track heading north through the sparse brush towards the first low ridge. To see the migmatites and associated metamorphic rocks, climb the small ridge to the first set of outcrops. The swirling effect of the lighter-colored granitic veinlets is apparent and a common direction can be traced along the ridge. Follow the rolling ridge crest about one-quarter mile further to the north. As you walk, notice that the same migmatitic rocks line the crest but just off the flanks near where the ridge begins to turn and merge with a more east-west oriented ridge, dark outcrops appear. These are amphibolites and represent ancient basalts injected into the older Julian Schist. Their melting temperature was higher than that attained during metamorphism and so they remained more or less intact.*

*R*eturn to your vehicle and continue ahead one mile (taking the left fork) to the end of the road at the overlook. This area is maintained by the BLM and has picnic tables plus historical information placards. The view, among the most spectacular in Southern California, extends past the Salton Sea to the Algodones sand dunes near Yuma. In winter and spring, the snow-capped San Jacinto Mountains are visible to the north. Early morning visitors may be treated to glimpses of desert bighorn sheep dancing up the steep slopes to the north and east. If you turn right at the fork just before the overlook, it will take you past the small spring that serves as a source of water for the animal life in this area. That road ends approximately one-half mile ahead.

From the overlook, the contact between the light-colored rocks of the La Posta pluton and the darker rocks of the roof pendant is clearly visible to the north. It swings westward behind the overlook area and then heads south, crossing the freeway just at the start of the downgrade into Jacumba. Looking eastward across the roof pendant, the tracks of the old San Diego and Arizona Eastern Railroad can be seen just across Carrizo Gorge winding their way northward around ridges and through small tunnels before disappearing behind the mountains to start its southward turn towards the Dos Cabezas siding (see Stop 2-1). This track system was built just after the turn of the century, mainly by Chinese laborers, and ran from San Diego through Campo and Jacumba Gorge and on into Arizona. Natural and man-made disasters to the tunnels and bridges in the gorge forced its closure in 1982.

The geology is as spectacular as the scenery. In all directions, but particularly to the south and east, whitish outcrops of **pegmatite dikes** (Figure 5.6) cut across the topography within the roof pendant. The gently sloping area immediately around the overlook contains numerous dikes up to nearly ten meters in thickness that parallel the ridge crest. The dikes are commonly layered with pegmatitic bands sandwiched between fine- to medium-grained granite. Both contain abundant feldspar and quartz with lesser amounts of whitish muscovite and/or dark, shiny biotite mica. Careful examination of

FIGURE 5.6 *View northeast from below Tule Mountain towards the Sacatone Springs overlook (middle right). The sinuous white streak cutting through the center of the photograph is a pegmatite dike that passes just below the overlook and continues on to the Pack Rat Mine just past a small gully. The dark rocks in the upper center are metamorphic rocks within the roof pendant.*

some outcrops, however, will reveal tiny, rounded, orange garnets and needles of black tourmaline (schorl). To the north along the ridge and across a small ravine, lies the sporadically active Pack Rat Mine and, out of sight beyond the top of the ridge, is the Beebe "hole." The latter produced a rumored $500,000 in gem spodumene after being discovered by a hunter, Lauren Beebe, in 1986. Although both properties are privately held and trespassing is discouraged, the area around the overlook and to the south have numerous exposures of the dikes. Small quarries along the west side of the ridge have excellent exposures of the internal structure of these pegmatites. These have been worked in the past for beryl, spodumene, tourmaline, garnet and other mineral specimens. Pockets, however, are rare and most specimens were encased in the rock.

Across the gorge to the east where the old S.D.A.&E. railroad tracks are visible, a buff-colored rock unit, the Indian Hill pluton, has punched its way into the reddish metamorphic rocks of the roof pendant (Figure 5.7). This body of rock is younger than the adjacent pegmatite dikes and cuts them off wherever the two are in contact. Studies of this pluton indicate that it was also produced by melting of the roof pendant rocks as they were engulfed by the rising La Posta magma.

A hike up the steep ridge to the northwest rewards the more adventurous with a spectacular 360° view and a series of outcrops that emphasize the enormous stresses that these ancient sedimentary rocks have undergone. The rocks are migmatites and quartz-biotite gneisses that have been severely folded during the uplift that accompanied the emplacement of the La Posta pluton which surrounds the roof pendant. Centimeter-thick layers of buff-colored quartz and feldspar contrast with the thin, dark layers of biotite and serve to highlight the tight folds within the rock. Studies of this rock unit indicate that it records at least two different periods of folding. The oldest deformation likely dates back to the forces that created the western zone of the PRB while the youngest measures the deformation caused by the rising La Posta pluton.

FIGURE 5.7 View to the northeast from the Sacatone Springs overlook. The remnants of the S.D.A.&E. track system cut around and through the buff-colored Indian Hill granite above Carrizo Gorge. The darker-colored rocks are migmatitic gneisses. See also Stop 2-2.

> One of the most striking aspects of this stop is the very sudden topographic drop down towards Carrizo Gorge.

A similar break in slope, seen at Stop 1-2, extends all along the eastern side of the Laguna Mountains and southward into Northern Baja California. This dramatic change may, in part, be due to the extensional forces that gripped Southern California and Northern Baja California during the Miocene (see Chapter 6). If so, faults of that age must occur parallel to the range fronts but they have not as yet been identified. The rugged character of the landscape and lack of road access is a deterrent to all but the most determined individuals.

Gold

The Julian District

 old is a mind-altering substance! Those who have never experienced the sight of a small, pea-sized nugget in a sluice box or watched a growing trail of fine gold dust collecting in a gold pan cannot understand the exhilaration and anticipation it instills in amateurs and professionals alike. I have panned nuggets from streams in Colorado and Northern California, and watched as, on two occasions in the Golden Chariot Mine, finely crushed rock washed over a simple table-like device left behind a gleaming trail of the magic powder. Each time, the excitement was unabated by previous experiences. Surely it must have been that way for the early prospectors in San Diego County.

Hints that Southern California had its share of gold can be found as far back as 1602 in the chronicles of Vizcaino's second expedition by Fra Antonio de la Ascencion. As quoted in Helen Ellsberg's book, *Mines of Julian*, he wrote that the beach sands contained ". . . great quantities of yellow pyrites, all full of holes—a sure sign that in the mountains . . . there are gold mines." The local Indians reportedly traded bird quills of gold dust with the Padres for supplies and later showed them the source of their yellow metal to work as their own. In 1849, an Army engineer named C. C. Parry, travelling down Banner Canyon towards the main road in San Felipe Valley, noted a change in rock type from "feldspathic or quartz granite" to "micaceous granite in which the scales of mica are frequently of large size and very confusedly intermixed" to "mica and talcose (talc-bearing) slates traversed by quartz veins." He concluded that "at this point, then, we have an approach to the gold formation." He had indeed crossed through the granitic rocks of the PRB, the Banner pegmatite district (Figure 5.3) and, lastly, the quartz veins within exposures of the Julian Schist that twenty years later would become the Julian gold district.

History

Gold was "officially" discovered in the Julian area in late 1869 by A. E. "Fred" Coleman, a veteran of the 1849 gold rush in Northern California who then lived near Volcan Mountain. His practiced eye caught a glint of the yellow metal in a small stream now known as Coleman Creek and he immediately fell to work using his fry pan as a gold pan. Within a few weeks, the Coleman Mining District had been formed and scores of miners lived in a loose collection of tents called Emily City, all drawn to the area by the promise of placer gold. Drawn to Southern California from the silver lodes in Nevada by rumors of work on an El Paso-to-San Diego Railroad came James and Drury Bailey and their cousins Mike and Webb Julian. Natives of Georgia, they had emigrated westward after the Civil War but found that those tragic years had left a residue of mistrust and hatred even as far away as San Diego. Unable to find work, apparently as a result of founder Alonzo Horton's policy of not hiring anyone with a Southern accent, they arrived at Emily City where a great many displaced Confederates had congregated.

Although their placer mining efforts were marginal, the cousins located the first lode deposit and on February 15, 1870, formed the Julian Mining District with their Warrior's Rest mine as the first recorded property. The Julian name was attached to the district by recorder Mike Julian but the townsite was named Julian by his cousin Drury Bailey who had acquired much of the land and was offering it free to anyone who would build homes or businesses there. When asked why "Julian" and not "Bailey," he is said to have replied that Julian, a family name, sounded better. Within a month, about twenty claims had been filed, all of which were in quartz veins similar to the Warrior's Rest and the hard rock mining history of the Julian District was underway.

Mining is a very risky proposition, subject to many and varied factors that can prevent a single mine or even a district from fulfilling its economic potential. The Julian District faced more than its share. A disputed boundary for the Rancho Cuyamaca land grant tied the independent miners up in court for several years and prevented them from directing their capital into improving the mines and milling operations around Julian. Although the miners eventually won their case in court, they were not able to recover from that financial blow, in part, because lending institutions in Southern and even Northern California refused to lend money to Southerners. At depth in the mines, the free-milling gold that characterized most of the district and made the extraction of the gold from the quartz ore so profitable, turned into a sulfide ore which could be efficiently processed only with expensive smelters. The mines fell into disrepair and many flooded as groundwater slowly filled the underground workings. Even the discovery of the district's richest lode deposit in 1871, the Golden Chariot mine, could not alter the inevitable and by the mid to late 1870s, the boom was over. There was a minor resurgence in 1888 with the discovery of the Gold King and Gold Queen mines just south of the main belt of mines in the district (Figure 5.8) but the easily extracted ore soon ran out and the boom gave way once again to bust. Today, independent miners or small partnerships still struggle to recover the precious metal from a few mines in the district but the glory days are long over.

Geology

*M*ost of the gold deposits that have been found in San Diego County occur within quartz veins that cut through an arcuate belt of the Julian Schist that runs from just north of Pine Valley through Julian and heads northwest towards Santa Isabel (Figure 2.1). Rocks to the east belong to the La Posta-type suite of igneous rocks whereas those to the west are gabbroic and granitic rocks of the western PRB. That this belt of gold-bearing rock sits in the medial portion of the PRB at the boundary between its western and eastern zones is no accident. Even a casual observer will recognize that the gold and gem pegmatite mining districts lie adjacent to one another and that both must then be related to the forces that created the eastern zone of the batholith.

Much of the Julian Schist in the area just southeast of the townsite and where the majority of mines occur consists of quartzite and fine-grained micaceous schist with lesser amounts of amphibolite, marble, and talc schist. These rocks, however, have a complex structural history in that they occur within or close to two fault systems (Figure 5.8). The Chariot Canyon fault zone runs south-southeast through the canyon for which it is named and appears to be the northwestern extension of the Cuyamaca-Laguna Mountain (CLM) fault zone that was discussed in an earlier section (see also Field Excursion Stops 1-3, 4). The Chariot Canyon fault zone intersects the active Elsinore fault zone near the bottom of Banner Grade. This much younger structural feature is related to the San Andreas fault system further to the east.

In Chariot Canyon, both the granitic rocks and the metamorphic rocks have been affected by at least two periods of movement, each of which has left a distinct imprint on the nature and distribution of the gold ore. Throughout the Chariot Canyon-CLM fault zone, the rocks have undergone a high-temperature shearing that resulted in a flow-like or plastic realignment of minerals into what geologists call mylonite. Superimposed on this is a younger,

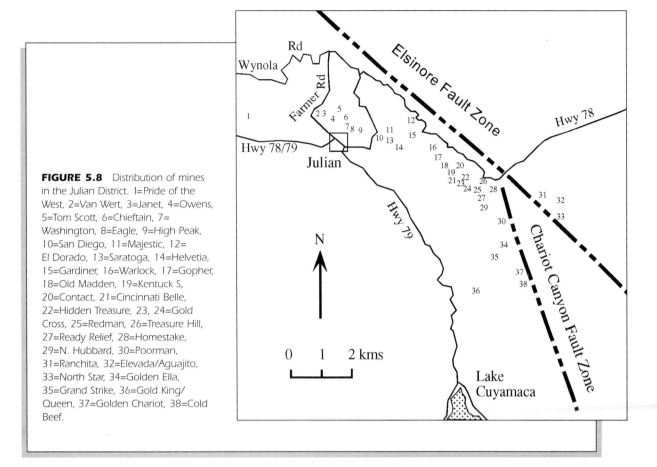

FIGURE 5.8 Distribution of mines in the Julian District. 1=Pride of the West, 2=Van Wert, 3=Janet, 4=Owens, 5=Tom Scott, 6=Chieftain, 7= Washington, 8=Eagle, 9=High Peak, 10=San Diego, 11=Majestic, 12= El Dorado, 13=Saratoga, 14=Helvetia, 15=Gardiner, 16=Warlock, 17=Gopher, 18=Old Madden, 19=Kentuck S, 20=Contact, 21=Cincinnati Belle, 22=Hidden Treasure, 23, 24=Gold Cross, 25=Redman, 26=Treasure Hill, 27=Ready Relief, 28=Homestake, 29=N. Hubbard, 30=Poorman, 31=Ranchita, 32=Elevada/Aguajito, 33=North Star, 34=Golden Ella, 35=Grand Strike, 36=Gold King/ Queen, 37=Golden Chariot, 38=Cold Beef.

low-temperature, crushing event that further reduced the grain size of what were then brittle rocks. The timing of both events is critical to our understanding of the development of the Julian gold field. As noted earlier, the main movement on the CLM shear zone (and presumably its northwestern extension in Chariot Canyon) must be older than 94 Ma (see Field Trip Stops 1-3,4). Its maximum age can be constrained by a 104 Ma tonalite pluton near the Cibbett Flat campground on Kitchen Creek Road, a portion of which has been affected by movement along the shear zone. Thus, movement along the CC-CLM must be younger than the age of the oldest affected pluton and older than that of the youngest unaffected pluton, i.e., between 104 and 94 Ma or about 100 Ma. It is this movement that produced the mylonitic or plastically deformed rocks in Chariot Canyon.

How does this constrain the age of the gold ores in the Julian District? Near Julian, the ores occur in tabular, dike-like bodies of quartz where, typically, the free-milling gold was a fraction of an ounce per ton of quartz. Moving into Chariot Canyon, these tabular quartz veins, a meter or so thick, become strung out into lenticular bodies much like a string of sausages (Figure 5.9). This form can only be produced by extensional or stretching forces that allowed the deeply buried, solid but still hot, tabular dikes to neck down into the series of lenticular forms observed today. This high-temperature stretching must correspond to movement along the CC-CLM zones and, since the gold was already in place by this time, indicates that the quartz veins and the gold mineralization must be no younger than about 100 Ma. Some of the vein quartz in the Golden Chariot and Cold Beef mines are reported to be highly fractured and crushed to the point that they were referred to as "sugar quartz" by the local miners. This granulation event is most likely related to a *reactivation* of the ancient CC-CLM shear zone by movement on the Elsinore Fault which has been active only over the last two million years. At that time, the rocks within the reactivated part of the Chariot Canyon Fault were near the surface and thus cooler and brittle.

The occurrence of gold within quartz veins is very common. Downward percolating groundwaters become heated if they come into contact with recently congealed magmas. These heated waters rise back towards the surface and preferentially dissolve silica and trace amounts of metals such as gold from the rock. As the rising hydrothermal solutions encounter cooler rocks near the surface, they precipitate the gold and quartz in fractures to form gold-bearing quartz dikes. This rather simple picture, however, does not readily fit the Julian District. If the gold-bearing quartz veins are at least 100 million years old, then the rocks that host them were buried at least 10 kilometers below the surface at the time of formation, much too deep for surface waters to penetrate. In addition, studies of minute inclusions of fluids trapped within the quartz vein and its associated minerals show that the temperature of the

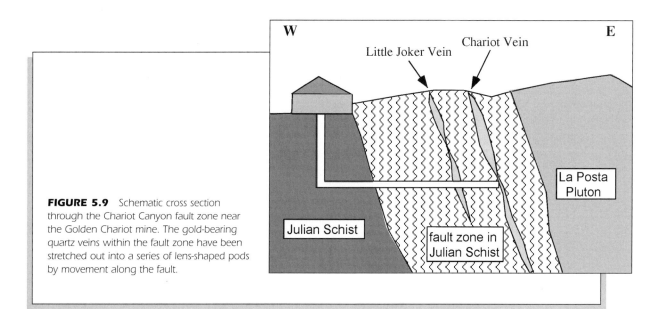

FIGURE 5.9 Schematic cross section through the Chariot Canyon fault zone near the Golden Chariot mine. The gold-bearing quartz veins within the fault zone have been stretched out into a series of lens-shaped pods by movement along the fault.

hydrothermal fluids was about 550°C (over 1000°F!!). These temperatures are corroborated by measurements of two isotopes of oxygen, O^{16} and O^{18}, that occur in pairs of minerals found associated with the dike. Their ratio, once measured in each of the two minerals, is dependent only on the temperature at which the minerals formed from the solution and yielded a calculated temperature of 531°, very close to that measured directly from the fluid inclusions. Fluids with these temperatures and oxygen isotope ratios could only come from metamorphic rocks that have been heated to the point that hydrous minerals such as mica break down and release the water bound within their structure. It is this ascending fluid, released from micas within deeply buried sections of the Julian Schist, that seems to have leached gold from ancient placers preserved within the quartzites. When the solution cooled to around 500°C, the gold precipitated along with the quartz to form the gold-bearing quartz veins exposed near Julian. Some quartz veins at places like the Stonewall mine near Cuyamaca Lake cut through granitic rock but are in close proximity to exposures of the Julian Schist and likely have a similar origin. The rocks from which the gold and the fluids were extracted must still be at considerable depth below the district.

Origin

*E*arlier in this chapter, it was pointed out that the Julian mining district and the belt of gem-bearing pegmatites sit side-by-side and astride the boundary between the western and eastern zones of the PRB. The time of emplacement of the gold veins is constrained only in that the veins must be older than about 100 Ma while limited age data on the pegmatites suggests that they formed between 95 and 100 Ma. This space-time relationship lends itself to a model that ties these very different systems together with the emplacement of the large La Posta pluton just to the east. The key to the problem lies in the recognition that this magma body was emplaced as a nearly crystal-free melt whose temperature, according to experimental studies, was at least 950°C. The temperature required to begin melting in the Julian Schist, i.e., that required for the micas to break down and release their structurally bound water is close to 650°C. It seems, then, that given the size of the pluton, its *heat content*, and the actual temperature of the La Posta melt when it intruded the Julian Schist, something had to happen.

Think of this as a two-stage process. As the rising La Posta melt approached the Julian Schist, muscovite, which contains 4–5% water, would be the first mineral to break down. When released at nearly 650°C, the water would migrate upwards through the schists and gneisses slowly leaching out soluble materials like gold and silica and redepositing them as gold-bearing quartz veins at cooler and shallower levels. As the melt fully engulfed the ancient metamorphosed sedimentary rocks, their temperatures would rise even higher and, in concert with the remaining released fluids, undergo partial melting to produce the pegmatitic magmas that became the gem-bearing dikes for which San Diego County is so famous. This explanation is firmly rooted in our current understanding of San Diego County's geologic history but awaits more scientific studies for confirmation.

Field Excursion 8

Gold

STOP 8-1

THE STONEWALL MINE. The Stonewall Mine and the remnants of its milling operations have been preserved as a historic site within Cuyamaca Rancho State Park. Visitors can wander around at their leisure and visit a small museum that depicts the glory days of gold mining in San Diego County. The Stonewall Mine was among the richest and most long-lived in the county and, although officially just outside the Julian Mining District, has much the same history and geology. The mine area can be reached via a well marked, paved park road that runs east from Highway 79 about two miles south of the Cuyamaca townsite in Cuyamaca Rancho State Park (Figure 2.3).

The quartz vein exploited in the Stonewall Mine was different from those in the Julian District in that it was consistent in thickness and cut through a granitic rather than schistose rock. Outcrops of schist, however, are nearby and it appears that this vein wandered from rock type to rock type. The ore grade was fairly steady throughout the vein which ranged up to 20 feet in thickness. In some areas where the vein appeared to thicken at a bend, gold values increased from an average of 1/2 ounce per ton to as much as 12 ounces per ton. At its peak operation between 1886–1892 under then California Governor Robert W. Waterman, the main shaft was extended to 600 feet, a thirty-stamp mill was in operation, hundreds of people were employed, and a total of about 2 million dollars in gold (at $20/oz.) was recovered. At today's prices, that amount of gold would bring in about 30 million dollars.

STOP 8-2

THE EAGLE-HIGH PEAK MINES. These two mines connect under High Peak on the northeast side of Julian and are currently operated as a tourist attraction. Signs in town direct interested visitors. This underground tour depicts mining processes, history, and working conditions during the heyday of the Julian District. Free-milling gold was extracted from three or four narrow quartz veins, each only a few inches thick. Gold production between 25 and 50 thousand dollars was reported from each operation during the life of the mining district.

STOP 8-3

THE CHARIOT CANYON TRUCK TRAIL. This Field Excursion starts at the intersection of Highways 78 and 79 in Julian (Figure 2.3). Proceed east on Highway 78 approximately 5.4 miles down Banner Grade. At this point, there is a wide dirt shoulder on the right (south) side of the road. From this vantage point, you can get a good view down Banner Canyon and out to Granite Mountain in the distance without crossing the road. Granite Mountain is underlain by La Posta-type granitic rocks whereas most of the exposures in Banner Canyon are metamorphic rocks mixed with minor amounts of western-zone granitic rocks. Banner Canyon carries the trace of the Elsinore Fault, a strike-slip fault that belongs to the San Andreas system. Look carefully along the far side of the canyon for several benches or subtle breaks in slope that can be traced for hundreds of yards. These are the topographic expression of different fault strands that together make up the Elsinore Fault zone. This portion of the fault zone has about two to three kilometers of right-lateral displacement, i.e., the opposite side of the fault has moved that distance to our right (southeast) relative to where you stand. Elsewhere, the Elsinore Fault has up to 30 kilometers of displacement. This means that there must be other fault strands in this area such as the one in Chariot Canyon that have taken up the additional movement.

*C*ontinue ahead another 2.5 miles to the unmarked right turn onto Chariot Canyon Road, about 100 yards north of the Banner Store. The road is ungraded and rough in places so that a high-centered vehicle is essential. Make this turn and drive about one-quarter mile past the old gate. Stop and enjoy the dramatic view back up Banner Canyon. From this point onward to the main collection of working mines about 2.4 miles ahead, the road passes through rather weathered outcrops of tonalite similar to that found in the La Posta pluton mixed with pegmatite dikes and large patches of schistose metamorphic rocks. Some of the pegmatite dikes terminate abruptly, probably along small faults that are difficult to see as a result of the deep weathering. The whitish surfaces on many of the road cuts are "caliche" deposits, i.e., calcite coatings on old fracture and fault surfaces. These form in arid weathering environments where rainfall dissolves soluble elements such as calcium and magnesium out of the rock. These elements combine with atmospheric carbon dioxide and are reprecipitated as "caliche" coatings as the solutions evaporate on the fracture surfaces. A drop or two of dilute hydrochloric acid will cause these coatings to effervesce.

At about 2 miles, the road reaches a high point where the headframe and buildings of the Golden Chariot and Cold Beef mines come into view (Figure 5.10). While all of this area is privately held and trespassing is forbidden, the road is public access. One-quarter mile ahead, a small adit just above the road drops vertically onto the Golden Chariot vein. Just ahead on the left before passing the entrance to the Golden Chariot Mine, a small outcrop of what may be part of the Golden Chariot vein is exposed. The outcrop has abundant quartz with small amounts of whitish material that may represent a clay alteration of feldspar. Note that part of the exposure is intensely fractured, probably as a result of more recent movement along the Chariot Canyon fault whose history goes back about 100 million years. Two-tenths of a mile further, the road passes the head frame of the Cold Beef mine and, after nearly another half-mile, the Freedom mill site at the Golden Oaks mine. Three-quarters of a mile ahead, the road enters Anza-Borrego Desert State Park and there are no further active mine sites.

From this point you can retrace your steps or continue ahead to see more spectacular geology as the road eventually winds its way down through Oriflamme Canyon to County Route S-2 just below Box Canyon. From the boundary sign for the State Park, drive for about one mile to a high point in the road. Most of the low, rolling terrane immediately around you is underlain by outcrops of the La Posta-type tonalite. These rocks, mapped by Cindi Lampe for her Masters Thesis in 1985, are similar to those described at Stop 6-1 and appear to be the outer portion of another La Posta-type pluton that has been disrupted by movement along the Elsinore Fault. To the southwest lies a ridge that forms the eastern edge of the Laguna Mountains. The road that climbs along this ridge is reached from a right turn about a one-fourth mile

FIGURE 5.10 View southeast along Chariot Canyon showing the structures and workings at the Golden Chariot mine site.

ahead. Follow that road for about one mile to the first series of road cuts that expose a well foliated mica schist. This metamorphic rock also contains small altered crystals of andalusite, a metamorphic mineral whose presence indicates that these rocks were buried to a maximum depth of about 10 or 11 kilometers during metamorphism. The andalusite commonly occurs as very small scattered grains that a practiced eye might find. However, several one- to two-foot thick layers along this section of the road contain abundant, well formed, fresh andalusite crystals up to about 5 mm across. Many of these are the chiastolite variety and have a dark, cross-like pattern that fits diagonally across the square cross section of individual grains.

If you drive about one mile further up this road, you come to a wide parking area on the left with a bit of stonework marking a spring house and well. The stonework bears the unmistakable signature of the Civilian Conservation Corps and the stop affords a cool, shady view back towards Chariot and Oriflamme Canyons. There is a locked gate a hundred yards or so ahead that marks the boundary to the Tulloch Ranch so this is a good place to turn around and retrace your drive back to the intersection with Chariot Canyon Road.

Turn right at the intersection for the spectacular drive down Oriflamme Canyon and through another segment of San Diego County geology. This road has been maintained and, as of early 1999, was passable to most vehicles. The steep drop into the canyon begins a little more than half a mile from the intersection and has a series of sharp wide switchbacks with ample parking space for one or two vehicles. The rocks exposed along these places are fault-related splinters of La Posta-type rocks that include both the Hornblende-Biotite Facies and the Large Biotite Facies. The metamorphic rocks occur either as inclusions within or as larger discrete bodies surrounded by the igneous rocks. Most are gneissic with light-colored swirls of quartz and feldspar alternating with dark bands of biotite similar to the migmatites observed at Stop 7-2 (see also Chapter 6). These rocks also contain scattered bits of garnet and sillimanite but not the andalusite seen earlier. This, coupled with other measurements of their depth of burial (geobarometry), indicate that this package of metamorphic rocks was buried perhaps as much as 3 or 4 kilometers deeper than the ones just a mile or two back. Since both packages of metamorphic rocks are at essentially the same elevation, those along Oriflamme Canyon must have been uplifted much more than those to the west and a long-dead ancient fault must lie somewhere in between. Since La Posta-type granitic rocks occupy the intervening area, it must have intruded along these faults and caused the piston-like uplift that would place the two metamorphic rock packages nearly side by side (see also Chapter 5).

Approximately three miles down Oriflamme Canyon, a left turn will take you through Rodriguez Canyon and back to Chariot Canyon Road near its intersection with Highway 78, a distance of about seven miles. If you continue straight ahead about two miles along Oriflamme Canyon Road, it will intersect County Route S-2. A left turn there will take you through Box Canyon and its excellent exposures of the same migmatitic rocks and on towards Scissors Crossing and Julian.

CHAPTER

6

THE MIOCENE

◿ Introduction

*A*t the start of Miocene time, San Diego County was relatively quiet and serene. The towering monoliths of the ancestral Peninsular Ranges Batholith had long since been eroded down to near their present elevations. A chain of low mountains ran continuously from the Sierras southward through San Diego County and along the western side of mainland Mexico. Peninsular California and the Gulf of California, however, were still millions of years into the future. Some of the familiar modern topography might even have been recognizable. Cuyamaca Peak and Los Pinos Mountain were likely beginning to emerge as positive erosional features. Ocean waters had retreated to near the present shoreline.

Elsewhere, major changes in plate configuration were taking place that would add a new dimension to the geology of San Diego County. In the area now occupied by Nevada and parts of southeastern California and western Arizona, the crust of the North American continent was being extended or stretched. Fault-outlined basins formed and began to fill with the eroded debris of the adjacent mountain blocks. Volcanic activity associated with this extension began during the Oligocene and continued into the early Miocene. Unlike the volcanism that occurs over subduction zones, these magmas were compositionally limited to rhyolite, the volcanic equivalent of a granite, and their eruptive style was very different. Large composite volcanoes built up of alternating lava flows and cinder cones are typical of subduction volcanoes. Here, however, extensive sheets of nearly solid, gas-rich magma called ignimbrites rolled out of fissures and spread out over vast areas on a low-relief erosional surface, burying everything in their paths.

Between 22 and 17 Ma, the early Miocene, this extension had reached into southeastern California. At about the same time, a small portion of the eastward subducting Farallon plate, the Monterey subplate, broke away beneath the western edge of North America and attached itself to the northeastward moving Pacific plate. This created a complex tectonic pattern in which the portion of the continent over the plate fragment began to twist clockwise. The middle portion of the Sierra Nevada-Peninsular Ranges mountain chain caught in this rotating block twisted clockwise and over the span of the next twenty

81

million years and underwent approximately 90° of rotation to form the appropriately named Transverse Ranges physiographic province (Figure 1.2). The rotation of this rigid continental block began to open a large sedimentary basin, the Los Angeles basin, as the crust was stretched and extended. As the continent thinned during the extensional process, it induced melting in the underlying mantle in much the same way that melting occurs beneath the extended mantle and oceanic crust at mid-ocean spreading centers (see Chapter 3). The result was a relatively short-lived period of volcanism within and around the Los Angeles basin. This activity appears to be restricted to between 13 and 16 Ma and left small volcanic fields scattered throughout Southern California from Conejo, the Palos Verdes peninsula, and Long Beach to as far south as Rosarito Beach in Baja California (Figure 6.1). This volcanism was different in form and composition from the rhyolitic ignimbrite sheets erupted a few million years earlier in what is now the Basin and Range Province (Figure 1.2). Small composite volcanoes, whose compositions ran the full gamut from basalt to rhyolite, dotted this landscape.

During early Miocene time, peninsular California was still attached to the North American continent but curious events seemed to foretell of the rifting that would eventually separate Baja from mainland Mexico. All along what is now the eastern edge of the Baja peninsula, scattered volcanic activity occurred from about 22 to 17 Ma. It was andesite in composition and very different from both the basalt to rhyolite packages centered about the Los Angeles basin and the rhyolitic ignimbrites of the Basin and Range. This volcanic activity appears also to have formed in an extensional environment but the tectonic controls for that setting are, as yet, uncertain.

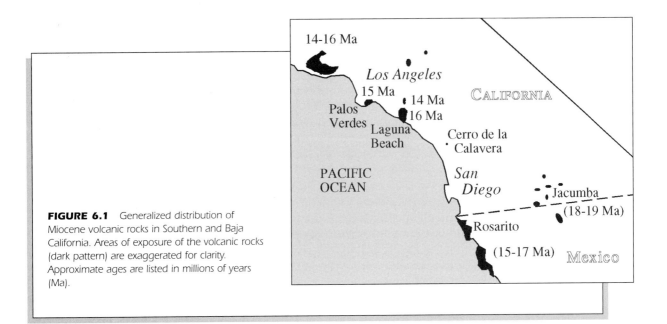

FIGURE 6.1 Generalized distribution of Miocene volcanic rocks in Southern and Baja California. Areas of exposure of the volcanic rocks (dark pattern) are exaggerated for clarity. Approximate ages are listed in millions of years (Ma).

△ Miocene Volcanism in San Diego County

*M*iocene volcanic rocks occur in several places within San Diego County. Its unique geographic position at the crude juncture of the three provinces described above lends itself to another geologic mystery. The main volcanic field in the county occurs near Jacumba, California, and is cut by both I-8 and Old Highway 80 (Figure 2.6). However, smaller volcanic features, presumably of Miocene age, can also be found at Cerro de la Calavera (Figure 6.2) near Carlsbad, Morro Hill at the southeastern edge of Camp Pendleton, and the black andesite dike that slides up onto Black's Beach, all located in coastal San Diego. Did they all form in the same tectonic framework or can we separate these volcanic materials on the basis of age and setting?

FIGURE 6.2 View of Cerro de la Calavera, a Miocene volcanic plug. An old stone quarry in the center of the structure exposes baked zones in Eocene sandstones intruded by the andesitic magma and near vertical columnar jointing.

The volcanic rocks around Jacumba are a mixture of lava flows and cinder cones that erupted about 18 million years ago. Some variation in composition was documented in a study in 1970 but the rocks effectively fall into the broad field of andesite. Thus, the timing, relatively restricted compositional range, and location at the northern end of the 17–22 Ma andesite province of Eastern Baja California suggests that these rocks belong to that province and owe their origin to whatever tectonic forces created that belt of volcanic centers.

The volcanic rocks exposed at Cerro de la Calavera and Morro Hill have no age data other than geologic field relationships that indicate they are younger than the late Eocene Torrey Sandstone. Field inspection indicates that these rocks are andesites but the occurrences are small and no detailed geologic work has been done on either body. Their proximity to the coast suggests that they are more likely related to the episode of volcanism initiated by the crustal rotations that reoriented the Transverse Ranges and created the Los Angeles basin.

Field Excursion 9

The Jacumba Volcanic Field

OVERVIEW. The Jacumba volcanic field lies in eastern San Diego County approximately 60 miles east of San Diego (Figure 2.6). Continue on I-8 from the College Ave. exit for about 52 miles until you pass the Boulevard exit. The freeway starts to drop from the high, rolling topography of the La Posta pluton (see Chapter 6) into Walker Canyon, a narrow cut through a mixed package of granitic and metamorphic rocks. About two miles further, the freeway bends to the southeast and begins to descend into Jacumba Valley.

As you cross the old S.D.A.&E. railroad tracks, the towering conical hill to your right (south) is Round Mountain, a small plug-like body of andesite. Exit just ahead at the Jacumba off-ramp and follow the road southeast about 1.1 miles to Old Highway 80. Turn left (east) and drive about one-half mile ahead to a wide shoulder on the north side of another large conical hill. Traffic is light here and it is a good place to park and get a general picture of this volcanic field.

To the north, you can see the flat top of Table Mountain (Figure 6.3) whose summit is capped by a series of nearly horizontal volcanic flows. Behind Table Mountain and out of your sight at this point is the highest mapped flow unit, the bottom of which is at an elevation of about 4,000 feet above sea level. Near the top the hill behind you, the nearly horizontal base of the same flow unit is at an elevation of about 3,400 feet above sea level, some 600 feet lower. Now look to the northeast towards the prominent bouldery ridge (Gray Mountain) west of Table Mountain. It is composed of granitic rocks of the Peninsular Ranges batholith and has a very sharp escarpment along its southwestern side (Figure 6.4). The lower topography left or west of that escarpment is underlain by more of the Jacumba Volcanics, including the same flow unit found

FIGURE 6.3 Table Mountain as seen from the southeast. The mesa-like top is capped by a thick flat-lying volcanic flow that is relatively resistant to erosion.

FIGURE 6.4 *Gray Mountain consists mainly of Cretaceous plutonic rocks of the Peninsular Ranges batholith. Its southwestern margin is a near vertical fault that brings the plutonic rocks into contact with Miocene volcanic rocks. The fault starts from the lower left side of the mountain, passes in front of the low hill in the center of the photo, and crosses Old Highway 80 to the viewer's right.*

on Table Mountain and the hill behind you. The base of the flow in this area, however, is at about 2,860 feet above sea level, nearly 1,000 feet below its counterpart on Table Mountain and more than 500 feet below the same flow on the hill behind you. Why such different elevations for the same flow?

Jacumba Valley is a zone of crustal extension or stretching. About 18 million years ago, the batholithic rocks a few miles to the east were being pulled to the northeast while those a few miles to the west were being pulled southwest. This extension was likely caused by a short-lived rise of hot plastic mantle rock beneath this portion of the crust which began to slide off the elevated mantle in several directions. This caused a series of steep northwest trending faults to form, one of which can be traced along the southwestern side of Gray Mountain, past the northeastern side of the hill where you are standing, and then southeastward into Mexico. Dozens of similar faults have been mapped in this valley by several geology field classes from San Diego State University. The northeast-southwest directed stretching created several downdropped fault blocks that give Jacumba Valley its topographic character and either shifted portions of the same flows to dramatically different elevations after solidification or caused them to flow over prominent fault scarps into valley bottoms.

ANDESITE PLUG. *Drive east again on Old Highway 80 about 1.5 miles to a dirt road that will go around a small hill mapped as an andesite plug similar to Round Mountain. Turn right (south) and drive around to the back side of the hill where outcrops of the andesite occur just above the road. Weathered phenocrysts (large crystals) of rectangular greenish pyroxene and amphibole are abundant along with scattered grayish plagioclase. This texture is referred to as porphyritic and indicates that the andesite melt underwent a period of slower cooling during which the phenocrysts grew, followed by a quenching of the melt to a fine-grained groundmass as it rose to surface levels. Reaction with groundwater over the past 18 million years has pref-*

erentially etched out the phenocrysts leaving behind iron oxide-coated voids in the shape of the original phenocrysts.

CINDER CONE. Continue following the dirt road around the hill until you reach Old Highway 80 again. Turn left (west) and go about one-quarter mile to another dirt road on the right (north) that turns west along the interstate. Follow this for one-half mile to an intersection. Turn right (north) and follow this road under the freeway towards Table Mountain. One-half mile ahead, the loading frame of the old Mica Gem pegmatite mine is visible. Turn left (west) on another dirt road about 100 meters before reaching the head frame. This road leads to a buried cinder cone and one of the basalt flows that cap Table Mountain. The first stretch is quite rough and passable only by four-wheel drive or high-centered vehicles. On a cool day, it is a comfortable walk of about one mile. As you drive or walk into the small quarry, note the layering that defines the shape of the cinder cone (Figure 6.5). To the north and 500 feet directly above the cinder cone, outcrops of the upper lava flow on Table Mountain are visible. The ridge west of the quarry drops down from Table Mountain and has outcrops of a similar flow that helped to bury the cinder cone. You can climb up to this flow and examine as fresh a sample as you can find, but broken pieces of the flow are abundant around the floor of the quarry. The flow contains scattered, oddly greenish, one- to two-millimeter phenocrysts of plagioclase set into a black, fine-grained matrix and suggests that the capping flow is an andesite (see Appendix A, Figure A-3). The other volcanic rocks above and thus younger than the cinder cone are a rather heterogeneous batch of tuff-breccias that likely represent lahars (volcanic mudflows) or pyroclastic flows that rumbled downslope from a vent higher on the mountain burying everything in their path.

*R*eturn to the dirt access road and head back towards I-8 (south). Pass under the freeway and proceed right (west) about 100 meters to a left (south) turn. Follow this dirt road under the power lines back to Old Highway 80.

FIGURE 6.5 Partly buried Miocene cinder cone exposed in an old quarry near Table Mountain. The layering near the center of the cone dips westward (left) but changes to near horizontal in the center of the photograph.

EPILOGUE

 Where Have We Been?

The igneous and metamorphic rocks in San Diego's back country contain the record of four magmatic events. The youngest (Miocene) is small in terms of both duration and quantity of new material added to the continent. It is important, however, because it is either part of the record of the complex subduction processes that rotated the Transverse Ranges into their current position and created the Los Angeles Basin, or it is a manifestation of an unknown event that strung out a series of volcanoes along the edge of what, 12 or 13 million years later, would become the Gulf of California. The other three magmatic events are related to the subduction of oceanic lithosphere beneath the leading (western) margin of the North American continent. These are important in that they are the record of the growth of the North American continent in the late Jurassic and Cretaceous and lend considerable insight into that process.

As one views the ancient rocks of the North American continent (Figure 7.1), a crude pattern emerges. The oldest ages are in the interior with successively younger rocks seemingly plastered onto the margins as slightly curved but elongate belts of igneous and metamorphic rock. These represent younger, subduction-related, volcanic/plutonic arcs that slowly merged with and were added to the older crust (Figure 3.3). Batholiths of the Sierra Nevada, the Coast Ranges of British Columbia, and the Peninsular Ranges were added to the continental margin during Jurassic and Cretaceous time (Figure 7.1) and represent the largest episode of continental growth in the Phanerozoic (see inside front cover). Thus, as we tracked through time following these events in San Diego County (and, by inference, the whole of Peninsular California), we watched as the North American continent added a vast expanse of newly formed rock to its margins. If not for this prolonged event, Tucson might have ocean-front properties.

Consider the granitic rocks of the Jurassic-Late Triassic CLM belt. They represent the plutonic roots of a volcanic arc that rose well above the sea floor as magmas derived from the melting of older, subducted, and altered oceanic crust moved upwards. They intruded into a series of sedimentary rocks, the precursors to the Julian Schist, and through the leading (western) edge of

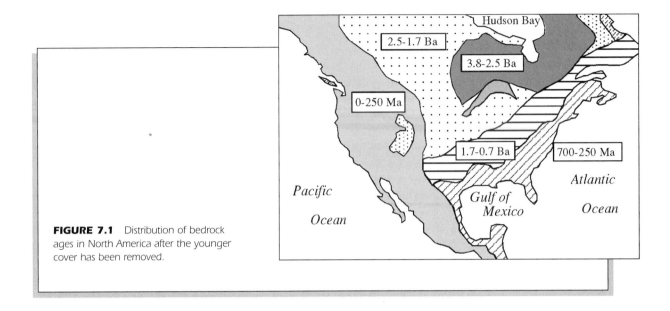

FIGURE 7.1 Distribution of bedrock ages in North America after the younger cover has been removed.

the North American continent. The arc was built up mainly on the edge of the continent and may have risen to great heights above the coastal plain. Rocks of similar composition and age are also found along the western side of the Sierra Nevada, in western Arizona, and in the northwestern portions of mainland Mexico. During Jurassic (and Late Triassic?) time, then, a sinuous magmatic arc, now deeply eroded to reveal its plutonic underpinnings, likely ran all along the western coast of North America crudely foretelling the eventual position of the younger (Cretaceous) Peninsular Ranges batholith.

What span of time separates these two events is not known because the number of radiometric ages in the older CLM arc is very limited. Are these really two separate events? The youngest CLM granite dated thus far is approximately 160 Ma whereas the oldest age for a PRB granitic rock is about 120 Ma. Are they related to the same, long-lived subduction event? Could a dramatic change in spreading rate and plate motion between 160 and 120 Ma have shifted magmatism elsewhere where the rocks lie hidden or unrecognized? Perhaps additional age data will show that there is no real time gap between them but until then, we must treat the two as separate events.

The main portion of the Peninsular Ranges is made up of three systems that overlap in time (Figure 1.1). The Santiago Peak Volcanics likely represents a volcanic arc that grew beneath and eventually emerged from the ocean outboard of both the North American continent and the older CLM magmatic arc. As it grew larger and more voluminous, magmas began to cool slowly deep below the volcanic edifices to form the plutons that are now exposed by erosion. The eventual uplift and erosion of this western PRB arc stripped the SPV cover off the top leaving only the scattered exposures now found in a discontinuous belt along the western edge of the exposed batholith. How these rocks are related to the plutonic rocks of the western zone of the batholith is still debated but the age constraints and contained rock types argue that they are part of the same geologic event.

The rocks of the western and eastern zones of the PRB represent two separate arcs that formed from two distinct plate configurations. The older western arc was created from a relatively steep subduction plane and consists of three distinct magma series (gabbro, tonalite, and monzogranite). Ages in greater San Diego County range between about 120 and 102 Ma. The monzogranites have some of the older radiometric ages but many appear to be younger than adjacent gabbroic plutons whose absolute age(s) are still in question. The gabbroic rocks form discrete bodies but appear to be related in space and time to the tonalitic plutons. Did the gabbroic melts form during the 160 to 120 Ma interval for which there are no known igneous bodies? Are only the youngest gabbroic plutons, those that rose to the highest levels in the crust, exposed at the surface with older rocks yet to be uncovered by erosion? Recent seismic studies of the crust beneath the western zone of the batholith indicate that it is much thicker (35–40 km) than expected. Is this the thickened gabbroic protocrust that underwent melting to create the tonalitic magmas?

At about 100 Ma, plate configuration changed. An increase in the spreading rate forced the subducted lithospheric slab to shallow and shifted the zone of melting eastward (Figure 4.18). Between about 98 and 94 Ma, great volumes of granitic melt rose to form the large La Posta-type plutons that characterize the eastern zone of the PRB. Why such large bodies of melt in such a short span of time? Although the answers are still being sought, the effects were of enormous benefit for Southern California. Heat convecting outward from the cooling plutons released water and other fluids held in the structure of micaceous minerals in the older Julian Schist. Some of the fluids, perhaps the earliest, leached precious metals such as gold trapped as ancient placers within the sandstones (now quartzites). As these fluids, whose temperatures reached at least as high as 570°C, cooled, the gold was redeposited in quartz veins in and around the shallower remains of that metamorphic system. As more fluids were released, the metasedimentary rocks underwent partial melting to create small amounts of water-rich granitic melts that rose into the older igneous and metamorphic rocks and cooled to form the gem-bearing pegmatites for which Southern California is so famous. The rise of massive amounts of magma into the upper portions of the asthenosphere between the older western arc and the edge of the North American continent did considerable damage to those older rocks. They were distorted and punched upwards into a mountain range that would certainly rival the Sierra Nevada today. So rapid was this uplift, at least on geological time scales, that erosion removed rock thicknesses of thousands of feet in a matter of a few tens of millions of years.

Where are these sediments that were derived from this uplift and erosion? It is not clear that they are anywhere within the local area. They do not appear to be in the modest amounts of Cretaceous sediments exposed along our coast. Are they offshore somewhere waiting to be found? Another fascinating suggestion argues, from paleomagnetic data, that these rocks have been displaced northwards all the way to British Columbia along now hidden strike-slip faults similar to the San Andreas. A Baja-B.C. connection? This idea still needs to be firmly established but is a convenient and mind-boggling explanation for the apparent disappearance of the eroded debris from the top of the Peninsular Ranges.

Where Are We Going?

The subduction process that created the Peninsular Ranges batholith is now quiet and inactive, at least in Southern California. Sometime after 90 Ma, the subduction plane shallowed again, eventually shifting the locus of magmatism far to the east. Between about 70 and 50 Ma, scattered bits of volcanic rock perforated the ancient continental crust in what is now central and western Arizona and southward into Mexico. In doing so, they provided the thermal energy required to create the massive copper deposits at Ajo, Globe, and Bagdad, and the fabled gold-silver veins at Tombstone. The volume of igneous rocks produced in this time period, however, was small compared to the events that gave rise to the Peninsular Ranges. The top of the subduction plane was simply too shallow and could not glean enough thermal energy from the asthenosphere to continue forming vast quantities of melt. To the west, the edge of the North American continent approached and, at about 30 Ma, eventually reached the spreading center that drove the subducted slab. The spreading center angled away from the continental plate so that this juncture slowly migrated northward creating a complex geologic picture. Once overridden by the continent, however, the spreading center's activity diminished and died. It now lies fossilized somewhere beneath the western portion of North American continent.

While the threat of subduction-related volcanism in Southern California is nil, our current geologic setting utilizes a different type of tectonics. Where two lithospheric plates, one containing oceanic lithosphere and one carrying a large continental "raft" (North America), had collided to create a series of magmatic arcs during Mesozoic time, the two plates now slip past one another along vertical faults. Driven by the northwestward movement of the Pacific plate, subduction gave way to lateral motion. The Peninsular Ranges physiographic province and the adjacent continental borderland are being sliced up and moved northwestward along a series of active strike-slip faults (Figure 7.2) whose surficial traces are dotted with and emphasized by earthquake epicenters. Many of these are described in Phil Kern's book *Earthquakes and Faults in San Diego County*. Strike-slip faults planes are essentially vertical but undergo horizontal movement such that the rocks on either side of the fault plane will move laterally past one another. They are further classified as right- or left-lateral, based on whether, relative to the observer, the opposite side of the fault has moved to the right or left. A quick examination shows that this classification is not dependent on the position of the observer. No matter what side of the fault is viewed, the designation of right- or left-lateral is the same.

In Southern California, including the greater San Diego area, the faults are all northwest-oriented and right lateral (Figure 7.2). If we stand on the eastern side of the San Andreas fault and look back to the west, everything within the Peninsular Ranges and Continental Borderland has been sliced into fault-bounded slivers that, as a package, are moving northwest. Each sliver has moved relative to the adjacent sliver such that the entire package behaves like a deck of cards lying on its side. Motion along the entire length of the San Andreas has averaged around 4 cm per year over the last 5 million years. Stepping west to the San Jacinto fault zone, slip rates of about 1.2 cm per year have been measured. The Elsinore Fault Zone varies between about 0.3 and

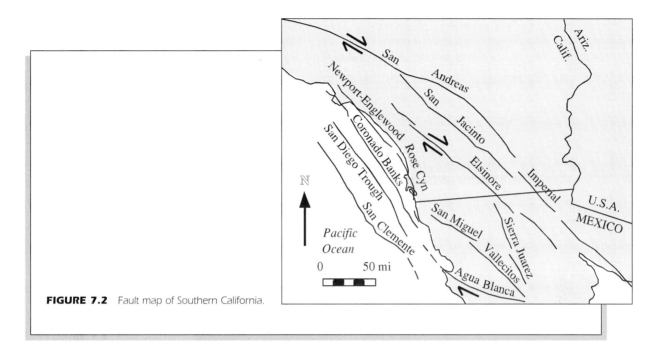

FIGURE 7.2 Fault map of Southern California.

0.5 cm per year along its entire length but it is the southern zone that currently is most seismically active. Both of these faults have been active for only about the last 4 or 5 million years.

The Rose Canyon Fault Zone is of most concern to the residents of the City of San Diego. Pieces of this fault system slice across the Silver Strand and merge with the main fault as it rides through Rose Canyon, and heads out to sea north of La Jolla. From here northward, the fault may merge with the Newport-Englewood fault zone for a total strike-length of approximately 200 km. Although slip rates for the entire system are a fraction of that of the San Andreas Fault Zone at 0.1 to 0.2 cm per year over the past several hundred thousand years, Richter magnitudes as high as 7.0 have been predicted for earthquakes on the Rose Canyon fault. To the south, the Agua Blanca fault slides at around 0.6 cm per year and threatens, with time, to shift the Baja Peninsula northwestward and isolate it from mainland Mexico. Offshore lie the Coronado Banks, San Diego Trough, and San Clemente fault zones with intervening splinters of Mesozoic crust. These faults appear to have been active only for the last 2 or 3 million years and have projected slip rates similar to their on-shore counterparts.

Using the current slip rate for the entire San Andreas system (3–5 cm or nearly 2 inches per year), a million years of motion will place fault-splintered San Diego County more than 30 miles further north. In 5 million years, we will be an island just north and west of the Los Angeles Basin. Will the northern part of Baja slip away from the rest of the peninsula along the Agua Blanca fault and join us on this journey? Will the Pacific flow in to join with the Gulf of California to create Isla Baja? Will new faults crack the thick crustal skin under the Peninsular Ranges and provide additional seismic hazards to future generations?

Despite all the unknowns, one thing is very clear. Nothing is permanent. Not mountain ranges, not continental configurations, not even the constellations in the night sky. The changes are slow, barely recognizable on human time scales, but they occur nonetheless. Cuyamaca Peak and Palomar Mountain will slide northwestward and eventually become beach sands on a distant shore. Burial, metamorphism, subduction, and remelting will continue the cycle as new mountains rise and new continents are shaped. And we are left to wonder.

APPENDIX A

CLASSIFICATION OF IGNEOUS ROCKS

The classification, i.e., naming, of any rock, igneous or otherwise, is always based on two parameters: composition and texture.

Composition can be given either as the chemical makeup of the rock, i.e., its percentage of common oxides such as SiO_2, Al_2O_3, TiO_2, FeO, MgO, CaO, Na_2O, and K_2O, or its mineralogical content. Of the two, mineralogical content is the most commonly used because it can be determined to a reasonable degree without the use of sophisticated and expensive analytical equipment.

Texture refers to the relationship between the individual minerals and involves estimates of grain size, both relative and absolute, grain shape, and grain orientation. It does not, as the name may imply, refer to the "feel" of a rock. Absolute grain size (the grain's measured dimensions) is important in that it allows for the distinction between plutonic and volcanic origins. If the rock is **phaneritic** or coarse grained, it means that the component grains are large enough to be seen with the naked eye or with a hands lens and formed through slow cooling and a period of extended growth deep within the crust. An **aphanitic** rock is fine grained such that, with the possible exception of a few larger crystals (**phenocrysts**), the individual grains cannot be discerned. This is the result of a very rapid cooling of the melt so that the individual minerals could not grow to any appreciable size. A **porphyritic** rock has two populations of grain sizes, one clearly larger than the other without regard to absolute size. Porphyritic volcanic rocks have an aphanitic matrix with phenocrysts whereas porphyritic plutonic rocks have a phaneritic matrix and distinctly larger phenocrysts. Both underwent two stages of cooling but at different levels within the Earth's crust.

The classification of igneous rocks is based on several triangular diagrams, each of which has a specific mineral or group of minerals (A, B, or C) at each apex. The apex represents 100% of that mineral. Any one side of the triangle can then be measured off in terms of the percentage of either of the two minerals at either apex. A point inside the triangle also must add up to 100% and gives the percentage of each of the three minerals.

The classification scheme for "granitic" igneous rocks is given in Figure A-1. These rocks contain at least two of the three framework minerals, quartz, plagioclase, and alkali feldspar and have a phaneritic (coarse-grained) texture. To "name" the rock, the percentages of the three framework minerals are estimated. Since there are other minerals such as biotite and hornblende in the rock, the three framework minerals will not add up to 100 and must be normalized to that value before they can be plotted on the classification diagram. In the example below, quartz, plagioclase, and alkali feldspar total 70% of the rock. When normalized, the rock falls into the granodiorite "pigeonhole" (Figure A-1). The rock name is then modified with the minerals *other than* those used to determine the "pigeonhole" name. Thus, the rock below becomes a muscovite-biotite granodiorite.

Composition	Normalized
21% quartz	21/.7 = 30% quartz
35% plagioclase	35/.7 = 50% plagioclase
14% alkali feldspar	14/.7 = 20% alkali feldspar
20% biotite	
10% muscovite	
100%	100%

The most difficult part of classifying granitic rocks is recognition of the three framework minerals, quartz, plagioclase feldspar, and alkali feldspar. Quartz is typically grayish and broken surfaces typically have an arcuate or partially rounded shape with a vitreous luster (the luster of broken glass). The

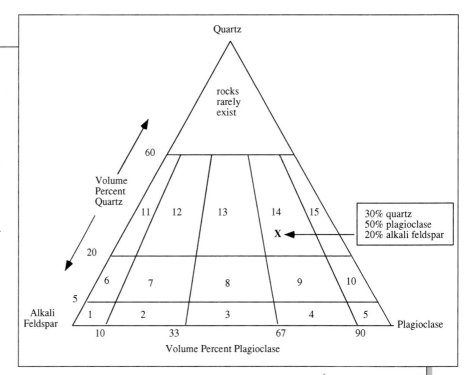

FIGURE A-1 Classification scheme for granitic rocks. numbers in "pigeonholes" refer to the following rock names: 1=alkali feldspar syenite, 2=syenite, 3=monzonite, 4=monzodiorite, 5=diorite, 6=alkali feldspar quartz syenite, 7=quartz syenite, 8=quartz monzonite, 9=quartz monzodiorite, 10=quartz diorite, 11=alkali feldspar granite, 12=syenogranite, 13=monzogranite, 14=granodiorite, and 15=tonalite.

two feldspars both have essentially identical shapes and two directions of cleavage at right angles (90°) to one another. In hand samples, their distinction is made with some difficulty. The feldspar family can undergo a phenomenon called *twinning* whereby, after the grain has formed and during the cooling process, segments of the crystal rotate slightly along a specific plane. Plagioclase tends to form multiple twin planes within single crystals, a phenomenon known as *polysynthetic twinning*. The twin planes are parallel to one of the cleavage directions but at nearly right angles to the other. Thus, when one examines the proper cleavage surface with a hand lens and under good lighting conditions, it appears to have a series of fine etched lines called twin striations. Once observed, this identifies the grain as plagioclase. Alkali feldspar, which comes in three chemically identical forms (sanidine, orthoclase, and microcline), has a tendency to produce no more than one of these twin planes that effectively bisects the grain into two halves. Thus, on careful examination, no more than one twin striation can be observed on a cleavage surface. This single twin plane, once observed, identifies alkali feldspar.

"Gabbroic" igneous rocks typically do not contain either alkali feldspar or quartz. They would all plot at a single point at the plagioclase apex in Figure A-1 and so require their own triangular diagram (Figure A-2). The apical minerals are plagioclase, olivine, and the sum of all other dark minerals, mainly pyroxene and amphibole. Normalization is not required because essentially all of the minerals are considered in the classification scheme. Mineral identification (and classification) is more easily done in these rocks because there is only one light-colored mineral, plagioclase. Olivine, pyroxenes, and amphiboles are dark. Fresh olivine commonly has a distinctive olive green color and a shiny (adamantine) luster. Pyroxenes and amphiboles are very difficult to distinguish from one another in hand samples and are simply lumped together in the classification scheme.

Other than the rocks made up of more-or-less one mineral species (dunite, anorthosite, pyroxenite, and hornblendite), the classification of gabbroic rocks depends on the proportion of plagioclase. If the rock has between 67 and 90% plagioclase with some combination of olivine, pyroxene, amphibole, etc., it is an anorthositic gabbro. Gabbro refers to rocks between 33 and 67% plagioclase whereas melanocratic (= dark colored) gabbro has 10–33%. Troctolite is an important biminerallic variety of gabbro that is made up of plagioclase and olivine. Peridotite, the rock that makes up most of the Earth's mantle, is composed of olivine and pyroxene.

Volcanic rocks are much more difficult to work with because you cannot see the bulk of the constituent minerals without the aid of a microscope or some other mineral identification equipment. Classification, by necessity, becomes very broad (Figure A-3) and is based on the type of phenocrysts (large crystals) observed. Biotite, hornblende, and other common minerals may occur as phenocrysts but the classification scheme is based only on the presence of the framework minerals, quartz, alkali feldspar, and plagioclase. Rhyolite, for example, contains phenocrysts of quartz and alkali feldspar but no plagioclase, whereas andesite has plagioclase phenocrysts but no quartz or alkali feldspar. **Basalt**, the volcanic equivalent of the gabbroic system, may have visible crystals of plagioclase and pyroxene but is recognized by the presence of olivine phenocrysts. The remainder of the rocks in the gabbroic system (e.g., peri-

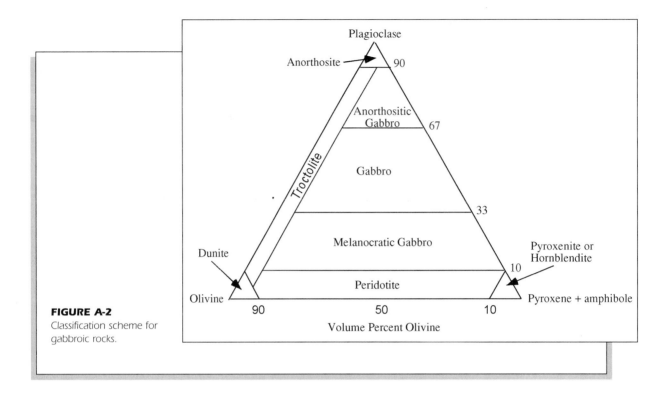

FIGURE A-2

Classification scheme for gabbroic rocks.

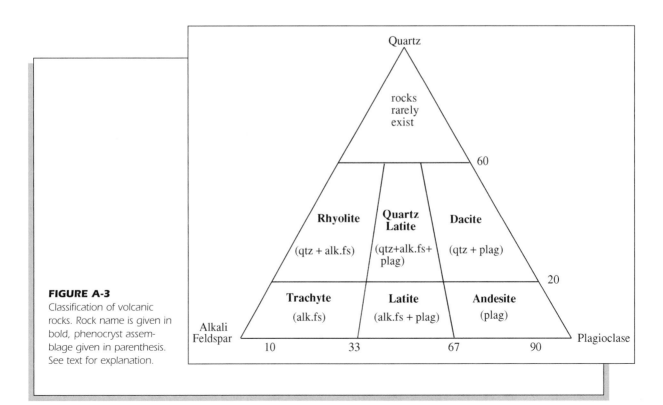

FIGURE A-3

Classification of volcanic rocks. Rock name is given in bold, phenocryst assemblage given in parenthesis. See text for explanation.

dotite, anorthosite, etc.) have no volcanic counterparts since these mono- or biminerallic rocks all form by crystal accumulation and do not represent a melt of that composition.

Fragmental volcanic rocks, i.e., those composed of pieces of other volcanic rocks, are formed when nearly solid magmas are exploded out of a volcanic center and mixed together with earlier materials. These rocks are classified on the basis of their fragmental texture and the size of the fragments. Loose material made up of fragments less than 4 mm in size is referred to as **ash** and the consolidated is **tuff**. Fragments between 4 and 256 mm are **lapilli** and a rock made up of this size fraction is a **lapillite**. Fragments greater than 256 mm are called **bombs** or **blocks** and a rock composed of these pieces is called a **volcanic breccia**. For rocks made up of a range of fragment sizes, a composite named is used. **Tuff-breccia**, for example, refers to a fragmental volcanic rock with a full range of fragment sizes (ash to bombs) whereas **tuff-lapillite** has a range from ash to lapilli (<4 to 256 mm).

APPENDIX B

CLASSIFICATION OF METAMORPHIC ROCKS

The classification scheme for metamorphic rocks is much less rigorous those for igneous rocks but again is based on both texture and composithantion. The characteristic feature of metamorphic rocks is an alignment of individual minerals that imparts either a **foliation**, a "leafy" appearance similar to the pages of a book, or a **lineation**, a statistical alignment of a few minerals into rod-like linear features. Coarse-grained rocks with a foliation are **schists** whereas those with a lineation are **gneisses**. Minerals such as biotite, muscovite, or hornblende that define the foliation or lineation are used as modifiers to the rock name, i.e., biotite schist or hornblende gneiss. Abundant minerals such as feldspar and quartz may be present in the rock but generally are not included in the rock name. However, low abundance metamorphic minerals such as sillimanite or staurolite that are important indicators of pressure and temperature are included whenever they appear. In addition to the above two rock names, other terms are used to designate specific coarse-grained metamorphic rocks that do not readily fit into the above two categories. Refer to the following table for these terms:

Quartzite	metamorphosed sandstone (mainly quartz with some feldspar).
Marble	metamorphosed limestone (calcite with lesser amounts of other calcium-bearing minerals such as tremolite, diopside, and grossularite garnet).
Skarn	metamorphic rock made up of some calcite along with significant amounts of other calcium-bearing silicate minerals such as tremolite, diopside, and grossularite garnet. This rock may either be the product of metamorphism of a clay-rich limestone or a result of reactions that occur between a marble and an intruding silicate magma.
Amphibolite	metamorphosed basalt (foliated hornblende-plagioclase rock).
Soapstone	metamorphosed olivine- and pyroxene-rich (peridotite?) rock now composed mostly of talc. The name is derived from the "soapy" feel of the rock.

Fine-grained metamorphic rocks are similar to volcanic (aphanitic) rocks in that their constituent minerals cannot be readily observed. This classification is mainly based on their form. Fine-grained metamorphic rocks with a distinct flat parting are **slate** whereas the same rock with a lustrous sheen to its surface is a **phyllite**. The sheen is caused by the presence of microscopic mica grains that grew during metamorphism. If these grains were large enough to be seen with the naked eye, the rock would become a schist.

APPENDIX C

MINERAL INDEX

Alkali feldspar: $(K,Na)AlSi_3O_8$. The variety of feldspar with potassium and sodium. It has three common forms with the same chemical composition—sanidine, orthoclase, and microcline. Sanidine is the high-temperature form that occurs only where quenched in volcanic rocks. Orthoclase and microcline occur in successively lower temperature plutonic rocks.

Amphibole: A complex family of silicate minerals with similar physical features and a general formula $A_{0-1}\,X_2Y_5\,(Si,Al)_8O_{22}(OH,F,CL)_2$ where A = Ca, Na, K; X = Ca, Fe, Mg, Mn, Na; Y = Al, Fe, Mg, Mn, Ti. Amphiboles form in both igneous and metamorphic environments. *See Hornblende.*

Andalusite: Al_2SiO_5. A metamorphic mineral formed by the reconstitution of clay minerals at low pressures and temperatures. *See also chiastolite.*

Aquamarine: The blue-green variety of the mineral beryl.

Beryl: $Be_3Al_2Si_6O_{18}$. A rare igneous mineral usually found in pegmatites. Its colored varieties include aquamarine (pale blue-green), emerald (dark green), heliodor (yellow), and morganite (pink).

Biotite: $K(Mg,Fe)_3(AlSi_3O_{10})(OH)_2$. The dark-colored, iron- and magnesium-bearing member of the mica family.

Calcite: $CaCO_3$. A common sedimentary mineral that forms by direct precipitation from aqueous solutions or through biological activity.

Chiastolite: The variety of andalusite with a diagnostic cross pattern of minute dark inclusions when viewed on end.

Chlorite: $(Mg,Fe,Al)_6Al,Si)_4O_{10}(OH)_8$. A metamorphic mineral that forms at relatively low pressures and temperatures. It can also form through hydrothermal alteration of other iron- and magnesium-bearing minerals.

Clay: A family of aluminum- and silica-rich minerals that commonly form by reaction of feldspars with hydrothermal solutions. *See also kaolinite.*

Cordierite: $(Fe,Mg)_2Al_4Si_5O_8$. A metamorphic mineral that is diagnostic of low pressures.

Diopside: $Ca(Fe,Mg)Si_2O_6$. A common igneous mineral and metamorphic member of the pyroxene family. As a metamorphic mineral, it forms at moderate to high pressures and temperatures.

Elbaite: The dark blue variety of tourmaline.

Epidote: $Ca_2(Al,Fe)_8Si_3O_{12}(OH)$. A common metamorphic mineral formed at low temperatures and pressures, or a mineral formed by the hydrothermal alteration of plagioclase feldspar.

Feldspar: A family of minerals consisting primarily of alkali feldspar and plagioclase. The most common mineral group in the Earth's crust.

Garnet: A family of minerals consisting primarily of grossularite ($Ca_4Al_3Si_4O_{12}$), pyrope ($Mg_4Al_3Si_4O_{12}$), and almandine ($Fe_4Al_3Si_4O_{12}$).

Hornblende: $NaCa_2(Mg,Fe,Al)_5(Al,Si)_8O_{22}(OH)_2$. The most common member of the amphibole family of minerals. It is found in both igneous and metamorphic rocks.

Ilmenite: $FeTiO_3$. An oxide of iron and titanium that forms under relatively reducing chemical conditions. *See also magnetite.*

Kaolinite: $Al_4Si_4O_{10}(OH)_8$. A member of the clay mineral group.

Kunzite: The pale purple variety of spodumene.

Magnetite: Fe_3O_4. An oxide of iron that forms under relatively oxidizing chemical conditions. See also ilmenite.

Mica: A family of minerals with a common sheet-like internal structure. *See also biotite and muscovite.*

Microcline: *See alkali feldspar.*

Morganite: *See beryl.*

Muscovite: $KAl_2(Al,Si_3)O_{10}(OH)_2$. The light colored member of the mica family.

Olivine: $(Mg,Fe)SiO_4$. A common igneous mineral in peridotite and basalt.

Orthoclase: *See alkali feldspar.*

Plagioclase: A member of the feldspar family consisting of a compositionally continuous series between albite ($NaAlSi_3O_8$) and anorthite ($CaAl_2Si_2O_8$).

Pyrite: FeS_2. Referred to as "fool's gold" because of its brassy color, it is a common accessory mineral in sulfide-rich ore deposits.

Pyroxene: A family of silicate minerals with a general formula $Y_2(Si,Al)_2O_6$ where Y = Ca, Na, Mg, Fe, Mn. Pyroxenes occur in both igneous and metamorphic rocks. *See also diopside.*

Quartz: SiO_2. The second most common mineral in the Earth's crust (after feldspar).

Rubellite: The red variety of tourmaline.

Sanidine: See alkali feldspar.

Schorl: The black variety of tourmaline.

Serpentine: $Mg_6Si_4O_{10}(OH)_8$. A product of the hydrothermal alteration of olivine and pyroxene.

Sillimanite: Al_2SiO_5. A metamorphic mineral formed by the recrystallization of clay minerals at moderate to high pressures and temperatures.

Sphene: $CaTiSiO_5$. A common accessory mineral in igneous rocks.

Spinel: $MgAl_2O_4$. An accessory mineral in some igneous rocks.

Spodumene: $LiAlSi_2O_6$. A rare mineral found in pegmatites. The gemmy green variety is hiddenite, the pink to violet variety is kunzite.

Talc: $Mg_3Si_4O_{10}$. A product of the hydrothermal alteration of olivine and pyroxene.

Tourmaline: $Na(Mg,Fe)_3Al_6(BO_3)_3(Si_6O_{18})(OH)_4$. A compositionally variable mineral found as an accessory in pegmatites and some hydrothermally altered rocks.

Vesuvianite (idocrase): $Ca_{10}Mg_2Al_4(Si_2O_7)_2(SiO_4)_5(OH)_4$. An accessory mineral found in metamorphosed limestones (marble).

Wollastonite: $CaSiO_3$. A metamorphic mineral associated with marble that forms under moderate temperatures and pressures.

Zircon: $ZrSiO_4$. An accessory mineral in igneous rocks that also contains enough uranium to make it useful in radiometric dating.

SELECTED READINGS

General Books

Earthquakes and Faults in San Diego County, by Philip J. Kern, 1993. Pickle Press, San Diego, CA. 90 pages.

Exploring and Mining Gems and Gold in the West, by Fred Rynerson, 1967. Naturegraph Publishers, Inc., Happy Camp, CA 96039. 204 pages.

Geology of Anza-Borrega, Edge of Creation, by Paul Remeika and Lowell Lindsay, 1992. Sunbelt Publications, 1250 Fayette St., El Cajon, CA 92020. 208 pages.

Geology of San Diego County, by Frederick W. Bergen, Harold J. Clifford, and Steven G. Spear, 1993. Sunbelt Publications, 1250 Fayette St., El Cajon, CA 92020. 175 pages.

Geology Underfoot in Southern California, by Robert P. Sharp and Allen F. Glazner, 1993. Mountain Press Publishing Company, Missoula, Montana. 224 pages.

Julian City and Cuyamaca Country, by Charles R. LeMenager, 1992. Eagle Peak Publishing Company, P.O. Box 1283, Ramona, CA 92065. 255 pages.

Mines of Julian, by Helen Ellsberg, 1972. La Siesta Press, P.O. Box 406, Glendale, CA 91209. 72 pages.

Geological Guidebooks for the Greater San Diego Area (Professional)

Earthquakes and Other Perils, San Diego Region. Patrick L. Abbott and William J. Elliot, eds., 1979. Guidebook for the Geological Society of America Field Trip. San Diego Association of Geologists, San Diego, CA. 226 pages.

Environmental Perils, San Diego Region. Patrick L. Abbott and William J. Elliot, eds., 1991. Guidebook for the Geological Society of America Field Trip. San Diego Association of Geologists, San Diego, CA. 250 pages.

Geological Excursions in Southern California and Mexico. Michael J. Walawender and Barry B. Hanan, eds., 1991. Field Trip Guidebook for the Geological Society of America. Department of Geological Sciences, San Diego State University, San Diego, CA. 515 pages.

Mesozoic Crystalline Rocks. Patrick L. Abbott and Victoria L. Todd, eds., 1979. Field Trip Guidebook for the Geological Society of America. Department of Geological Sciences, San Diego State University, San Diego, CA. 515 pages.

Reference Texts in Geology

Dynamic Earth, by Brian J. Skinner and Stephen C. Porter, 1995. John Wiley and Sons, Inc., New York. 567 pages.

Manual of Mineralogy, by Cornelius Klein and Cornelius Hurlbut, 1998. 21st edition, John Wiley and Sons, Inc., New York. 704 pages.

Mineralogy, by L. G. Berry, B. Mason, and R. V. Deitrich, 1983. W. H. Freeman and Company, San Francisco, CA. 561 pages.

Petrology, by Loren A. Raymond, 1995. Wm. C. Brown Publishers, Dubuque, IA. 742 pages.

Physical Geology, by Sheldon Judson and Maurice Kauffman, 1990. Prentice-Hall, Englewood Cliffs, NJ. 534 pages.

Understanding the Earth, by Frank Press and Raymond Vine, 1994. W. H. Freeman and Co., New York. 593 pages.

Selected Masters Theses
(Available at Love Library, S.D.S.U.)

Adams, M. A., 1979, Stratigraphy and Petrography of the Santiago Peak Volcanics East of Rancho Santa Fe, California. 123 pages.

Berggreen, R. G., 1976, Petrography, Structure, and Metamorphic History of a Metasedimentary Roof Pendant in the Peninsular Ranges Batholith, San Diego, California. 77 pages.

Clinkenbeard, J. C., 1987, Mineralogy, Geochemistry, and Geochronology of the La Posta Pluton, San Diego and Imperial Counties, California. 215 pages.

Corollo, G. F., 1993, Geochemistry, Petrography, and REE Analyses of Leuco-granite Plutons from the Western Zone of the Peninsular Ranges Batholith. 177 pages.

Eastman, B. G., 1986, The Geology, Petrography, and Geochemistry of the Sierra San Pedro Martir Pluton, Baja California, Mexico. 153 pages.

Foster, B. D., 1994, Origin and Tectonic Significance of Peninsular Ranges Amphibolites. 112 pages.

Germinario, M. P., 1982, The Depositional and Tectonic Environments of the Julian Schist, Julian, California. 95 pages.

Gorzolla, Y. P., Geochemistry and Petrography of the Santiago Peak Volcanics, Santa Margarita and Santa Ana Mountains, Southern California. 95 pages.

Hoffman, D. R., 1976, Petrogenesis of the Lawson Peak Orbicular Gabbro, San Diego, California. 95 pages.

Hoppler, H., 1983, Petrology and Emplacement of the Long Potrero Pluton: A Tail of One Pluton. 45 pages.

Kimsey, J. A., 1982, Petrography and Emplacement of the La Posta Granodiorite. 78 pages.

Kofron, R. J., 1984, Age and Origin of Gold Mineralization in the Southern Portion of the Julian Mining District, Southern California. 75 pages.

Lampe, C. M., 1988, Geology of the Granite Mountain Area; Implications for the Extent and Style of Deformation along the Southeastern Portion of the Elsinore Fault. 150 pages.

Lillis, P. G., 1978, Petrography, Geochemistry, and Structure of the Corte Madera Pluton, San Diego County, California. 80 pages.

Parrish, K. E., 1990, Geology, Petrology, Geochemistry and Isotopic Character of the Indian Hill Pluton, Jacumba Mountains, California. 136 pages.

Wernicke, R. S., 1987, Geology, Petrography, and Geochemistry of the El Topo Pluton, Baja California Norte, Mexico. 226 pages.

GLOSSARY

Absolute age: Age as measured in a specific quantity such as years, millions of years (Ma), or billions of years (Ba) before present. *See also relative time.*

Ammonite: An extinct mollusk related to the modern chambered nautilus.

Asthenosphere: The weak layer in the Earth's mantle that sits below the lithosphere. It deforms plastically when stressed and is the layer in the Earth on which the lithospheric plates ride.

Batholith: An exposed, irregular mass of igneous rock greater than 100 km² in total area and composed of individual plutons.

Billion: 10^9 or one thousand million. 1,000,000,000.

Caliche: A coating of the mineral calcite ($CaCO_3$) on rock fracture surfaces deposited from the percolation of groundwater over the fracture surface.

Chert: A sedimentary rock composed of microcrystalline quartz.

Colluvium: Loose weathered material that has accumulated on slopes below a topographic high.

Comb layering: A type of mineral layering that is similar to laying a series of combs next to one another with the "teeth" all pointing in the same direction and the ends of the "teeth" butting against the back of adjacent combs.

Conglomerate: A sedimentary rock composed predominantly of mineral particles or rock fragments that are greater than 2 mm in size.

Cross beds: A stacked sequence of variably inclined beds in a sedimentary rock.

Crystallization differentiation: An internal process in a magma chamber that separates early formed crystals from the parent melt, thus changing the composition of the remaining (daughter) melt.

Deformation: Any process by which preexisting rocks are bent or broken, or their component minerals are reoriented.

Dike: A planar or tabular mass of igneous rock that formed by the injection of melt into a preexisting fracture that cuts across older rock structures.

Flysch: A package of sedimentary rock that consists of alternating layers of sandstone and shale.

Fold: An aspect of rock deformation in which planar features are bent into a series of wrinkles. As measured from crest to crest, folded rock layers may have amplitudes that range from millimeters to kilometers across.

Foliation: A leafy appearance in a metamorphic rock similar to the pages of a book. It is caused by the realignment of platy minerals such as mica during metamorphism.

Forearc basin: In a plate tectonic setting, it is the basin or topographic depression found between a volcanic arc and its associated trough.

Geochronology: The branch of the geological sciences that deals with the age of rocks.

Geomorphic: An adjective that refers to the topographic shape of landforms.

Geothermal: Involving heat from within the Earth.

Gneiss: A metamorphic rock with elongate clusters of dark minerals. See Appendix B.

Gruss: Deeply weathered, disintegrated igneous rock.

Hydrothermal: Pertaining to geothermally heated waters.

Igneous: A category of rock formed by the crystallization of molten material. *See also magma.*

Ignimbrite: A volcanic rock that erupts as a gas-charged, nearly solid flow.

Inclusion: A fragment of older rock encased in an igneous rock. *See also xenolith.*

Intra-plate hot spot: A limited region within a tectonic plate that consistently erupts volcanic materials.

Intrusion: Molten material that has forced its way into preexisting rocks and solidified. *See also pluton.*

Island arc: An arcuate belt of volcanic islands that marks the subduction of one lithospheric plate beneath another. *See also volcanic arc.*

Isotope: One form of an element that has the same number of protons in the nucleus but a different number of neutrons and thus a different atomic weight from the other forms of that element.

Lahar: A debris flow consisting of newly deposited volcanic material mixed with rainwater.

Left lateral: A classification of strike-slips faults in which the opposite side of the fault moves to the observer's left. *See also right lateral.*

Limestone: A sedimentary rock consisting primarily of calcite.

Lithosphere: The outer, rigid shell of the Earth that consists of continental and oceanic crust as well as approximately 80–100 km of underlying mantle.

Magma: Molten rock material consisting of melt and any contained solids.

Mantle: That portion of the Earth that lies between the crust and the Earth's outer core.

Mass Wasting: Downslope movement of soil and rock under the influence of gravity.

Melange: A heterogeneous mixture of rock materials typically found near the boundary between two colliding tectonic plates.

Melt: The molten portion of a magma.

Metaconglomerate: A metamorphosed conglomerate.

Metasediment: Any metamorphosed sedimentary rock.

Mid-ocean spreading center: A linear belt along the sea floor where two lithospheric plates are being split apart and new material in the form of basaltic magma is rising.

Migmatite: A mixed rock formed by partial melting of a preexisting metamorphic rock. It contains both the solidified molten material and the refractory or unmelted residue of the original rock.

Mudstone: A fine-grained sedimentary rock that consists primarily of particles less than 0.06 mm in size.

Pegmatite: A very coarse-grained igneous rock with an average grain size exceeding 3 cm.

Physiographic province: An extensive package of rocks that have a similar geologic history and lie within a recognizable geomorphic setting.

Placer deposit: A deposit of fragmental sedimentary material that contains heavy metallic minerals such as gold near its base.

Pluton: An intrusion of igneous material that solidified at depth. See also batholith.

Porphyry: An igneous rock with two populations of grain sizes, one distinctly larger than the other.

Pressure: The weight of a column of rock overlying a point within the Earth. One kilobar of pressure is equal to about 3.5 km (2.1 miles) depth.

Protolith: The source rock from which a given igneous, metamorphic, or sedimentary rock was formed.

Radioactive decay: A process whereby the isotopes of certain elements spontaneously emit either subatomic particles or energy and create a new element or an isotope of the same element.

Relative age: Age as measured relative to, i.e., before or after, other events.

Rifting: Spreading apart or extension of a given area.

Right lateral: A classification of strike-slips faults in which the opposite side of the fault moves to the observer's right. *See also left lateral.*

Ripple marks: Tiny cusp-like ridges left in sands or muds as a result of oscillatory movement of water or wind.

Roof pendant: A large segment of metamorphic rock completely encased in a younger igneous rock.

Sandstone: A medium-grained sedimentary rock that consists primarily of particles between 0.06 and 2.0 mm in size.

Schist: A metamorphic rock characterized by a well developed foliation.

Sedimentary rock: A rock formed by the accumulation and lithification of weathered debris.

Sill: A planar or tabular mass of igneous rock that formed by the injection of melt into a preexisting fracture that is parallel to older rock structures.

Skarn: A metamorphic rock consisting of calcium- and magnesium-rich silicate minerals such as diopside, wollastonite, and grossularite garnet.

Stratigraphy: Pertaining to the stacking or sequential deposition of sedimentary rocks.

Trench: A lateral topographic depression that marks the subduction of one lithospheric plate beneath another.

Vesicles: Small, round, open spaces in volcanic rocks created by the melt solidifying around water bubbles.

Volatile: Any substance which, at surface pressures but magmatic temperatures, is a gas.

Volcanic arc: An arcuate belt of volcanic islands that marks the subduction of one lithospheric plate beneath another. *See also island arc.*

Xenolith: A fragment of older rock encased in an igneous rock. *See also inclusion.*

INDEX